T0275793

CAMBRIDGE MONOGRAPHS ON MATHEMATICAL PHYSICS

General editors: P. V. Landshoff, D. W. Sciama, S. Weinberg

FUNCTIONAL INTEGRALS AND COLLECTIVE EXCITATIONS

FUNCTIONAL INTEGRALS AND COLLECTIVE EXCITATIONS

V. N. POPOV

Leningrad Branch of V. A. Steklov Mathematical Institute of the Academy of Sciences of the USSR

The right of the University of Cambridge to print and sell all manner of books was granted by Henry VIII in 1534. The University has printed and published continuously since 1584.

CAMBRIDGE UNIVERSITY PRESS

Cambridge New York Port Chester Melbourne Sydney

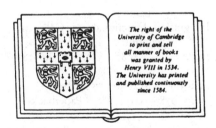

CAMBRIDGE UNIVERSITY PRESS
Cambridge, New York, Melbourne, Madrid, Cape Town, Singapore,
São Paulo, Delhi, Dubai, Tokyo, Mexico City

Cambridge University Press
The Edinburgh Building, Cambridge CB2 8RU, UK

Published in the United States of America by
Cambridge University Press, New York

www.cambridge.org
Information on this title: www.cambridge.org/9780521407878

© Cambridge University Press 1987

This publication is in copyright. Subject to statutory exception
and to the provisions of relevant collective licensing agreements,
no reproduction of any part may take place without the written
permission of Cambridge University Press.

First published 1987
First paperback edition (with corrections) 1990

A catalogue record for this publication is available from the British Library

Library of Congress Cataloguing in Publication Data

Popov, V. N. (Viktor Nikolaevich)
Functional integrals and collective excitations.
"Partly based on my lectures at the Institut für
Theoretische Physik, Freie Universitat, West
Berlin" – Pref.
Bibliography
Includes index.
1. Integration, Functional. 2. Collective
excitations. I. Title.
QC20.7.F85P65 1987 530.1′55 86–28398

ISBN 978-0-521-30777-2 Hardback
ISBN 978-0-521-40787-8 Paperback

Cambridge University Press has no responsibility for the persistence or
accuracy of URLs for external or third-party internet websites referred to in
this publication, and does not guarantee that any content on such websites is,
or will remain, accurate or appropriate. Information regarding prices, travel
timetables, and other factual information given in this work is correct at
the time of first printing but Cambridge University Press does not guarantee
the accuracy of such information thereafter.

Contents

Preface

Nowadays functional integrals are used in various branches of theoretical physics, and may be regarded as an 'integral calculus' of modern physics. Solutions of differential or functional equations arising in diffusion theory, quantum mechanics, quantum field theory and quantum statistical mechanics can be written in the form of functional integrals.

Functional integral methods are widely applied in quantum field theory, especially in gauge fields. There exist numerous interesting applications of functional integrals to the study of infrared and ultraviolet asymptotic behaviour of Green's functions in quantum field theory and also to the theory of extended objects (vortex-like excitations, solitons, instantons).

In statistical physics functional methods are very useful in problems dealing with collective modes (long-wave phonons and quantum vortices in superfluids and superconductors, plasma oscillations in systems of charged particles, collective modes in ^{3}He-type systems and so on).

This book is devoted to some applications of functional integrals for describing collective excitations in statistical physics. The main idea is to go in the functional integral from the initial variables to some new fields corresponding to 'collective' degrees of freedom. The choice of specific examples is to a large extent determined by the scientific interests of the author.

We dwell on modifications of the functional integral scheme developed for the description of collective modes, such as longwave phonons and quantum vortices in superfluids, superconductors and ^{3}He, plasma oscillations in systems of charged particles. The last chapters are devoted to the functional integral approach to the theory of crystals, the theory of many electron atoms and also to the theory of some model systems of statistical physics, exactly integrable in the sense of N. N. Bogoliubov (the BCS model and the Dicke model).

Problems of rigorous justification of functional integral formalism are almost not touched upon in this book. Here functional methods are used as a powerful tool to build up perturbation theory or to go over from one perturbative scheme to another. The only exclusion is the last chapter devoted to the BCS and Dicke models. Here the results concerning the Dicke model are obtained in the mathematically rigorous way. These results, up to now, are more strong than that obtained by any other methods.

This book is partly based on my lectures at the Institut für Theoretische Physik, Freie Universität, West Berlin, FRG (Popov, 1978).

I am very grateful to A. G. Reiman, for many remarks on the style of the book.

Part I

Functional integrals and diagram techniques in statistical physics

1
Functional integrals in statistical physics

Functional integration is one of the most powerful methods of modern theoretical physics. The functional integration approach to systems with an infinite number of degrees of freedom turns out to be very suitable for the introduction and formulation of the diagram perturbation theory in quantum field theory and statistical physics. This approach is simpler than that using an operator method.

The application of functional integrals in statistical physics allows one to derive numerous interesting results more quickly than when using other methods. The theories of phase transitions of the second kind, superfluidity, superconductivity, lasers, plasma, the Kondo effect, the Ising model – a list of problems which is far from complete, for which the application of the functional integration method appears to be very fruitful. In some of the problems, it allows us to give a rigorous proof of results obtained by other methods. If there is a possibility of an exact solution, the functional integration method gives a simple way to obtain it. In problems far from being exactly solvable (for example, the general theory of phase transitions), the application of functional integrals helps to build up a qualitative picture of the phenomenon and to develop the approximative methods of calculations.

Functional integrals are especially useful for the description of collective excitations, such as plasma oscillations in the system of particles with Coulomb interaction, quantum vortices and long-wave phonons in superfluidity and superconductivity, collective modes in ^{3}He-type systems and in ^{3}He–^{4}He mixtures. That is the case when standard perturbation schemes should be modified. Functional integrals represent a sufficiently flexible mathematical apparatus for such a modification.

In what follows the reader will find the main information on functional integrals in quantum statistical physics describing systems of interacting particles at finite temperature T. The functional integration method is suitable for obtaining the diagram techniques of the perturbation theory, and also for modifying the perturbative scheme if such a modification is necessary (for example, superfluidity, superconductivity).

There exist several modifications of Green's functions for quantum mechanical systems at nonzero temperatures – temperature, temperature–time Green's functions and so on. The diagram techniques can be directly derived only for temperature (Matsubara, 1955) Green's functions. They are convenient for calculating the thermodynamical properties of the system. They also contain information on the quasiparticle spectrum and weakly nonequilibrium kinetical phenomena.

The theory of temperature Green's functions in the operator approach can be found in an excellent book (Abrikosov, Gor'kov and Dzyaloshinki, 1962).

2

Functional integrals and diagram techniques for Bose particles

Let us consider a system of Bose particles at a finite absolute temperature T in a large cubic volume $V = L^3$. Here and below we will use the system of units with $\hbar = k_B = 1$ ($\hbar$ is the Planck constant divided by 2π and k_B is the Boltzmann constant).

It turns out that we can associate with any Bose system a functional integral over the space of complex functions $\psi(\mathbf{x}, \tau)$ ($\mathbf{x} \in V, \tau \in [0, \beta]$, $\beta = T^{-1}$).

We require $\psi(\mathbf{x}, \tau)$ to be a periodic function of the variable τ, so that $\psi(\mathbf{x}, \beta) = \psi(\mathbf{x}, 0)$. In addition, it is suitable to impose periodic boundary conditions with respect to the $\mathbf{x}$ variable. We may thus expand $\psi(\mathbf{x}, \tau)$, $\bar{\psi}(\mathbf{x}, \tau)$ into Fourier series

$$\psi(\mathbf{x}, \tau) = (\beta V)^{-1/2} \sum_{\mathbf{k}, \omega} e^{i(\omega\tau + \mathbf{k}\mathbf{x})} a(\mathbf{k}, \omega),$$

$$\bar{\psi}(\mathbf{x}, \tau) = (\beta V)^{-1/2} \sum_{\mathbf{k}, \omega} e^{-i(\omega\tau + \mathbf{k}\mathbf{x})} a^*(\mathbf{k}, \omega). \tag{2.1}$$

Here $a(\mathbf{k}, \omega)$, $a^*(\mathbf{k}, \omega)$ are the Fourier coefficients,

$$\omega = 2\pi n/\beta, k_i = 2\pi n_i/L, n, n_1, n_2, n_3 \text{ are integers.} \tag{2.2}$$

Let us introduce the functional

$$S = \int_0^\beta d\tau \int d^3x \bar{\psi}(\mathbf{x}, \tau) \partial_\tau \psi(\mathbf{x}, \tau) - \int_0^\beta H'(\tau) d\tau, \tag{2.3}$$

which plays the role of the action functional for the system. Here $H'(\tau)$ is the so-called generalized Hamiltonian,

$$H'(\tau) = \int d^3x \left(\frac{1}{2m} \nabla \bar{\psi}(\mathbf{x}, \tau) \nabla \psi(\mathbf{x}, \tau) - \lambda \bar{\psi}(\mathbf{x}, \tau) \psi(\mathbf{x}, \tau) \right)$$
$$+ \frac{1}{2} \int d^3x \, d^3y \, u(\mathbf{x} - \mathbf{y}) \bar{\psi}(\mathbf{x}, \tau) \bar{\psi}(\mathbf{y}, \tau) \psi(\mathbf{y}, \tau) \psi(\mathbf{x}, \tau), \tag{2.4}$$

where $u(\mathbf{x} - \mathbf{y})$ is the pair potential function and λ is the chemical potential coefficient.

Now we can define the one-particle Green function of the system by

$$G(\mathbf{x},\tau;\mathbf{y},\tau') = -\langle\psi(\mathbf{x},\tau)\bar{\psi}(\mathbf{x}',\tau')\rangle, \tag{2.5}$$

where

$$\langle\psi(\mathbf{x},\tau)\bar{\psi}(\mathbf{x}',\tau')\rangle = \frac{\int e^S\psi(\mathbf{x},\tau)\bar{\psi}(\mathbf{x}',\tau')\mathrm{D}\bar{\psi}\mathrm{D}\psi}{\int e^S\mathrm{D}\bar{\psi}\mathrm{D}\psi} \tag{2.6}$$

is a (formal) quotient of two functional integrals over a space of complex-valued functions. The measure of integration is denoted by $\mathrm{D}\bar{\psi}\mathrm{D}\psi$.

It is convenient to use the Fourier decomposition (2.1) and rewrite the functional S in terms of the Fourier coefficients

$$S = S_0 + S_1, \tag{2.7}$$

where

$$S_0 = \sum_p \left(i\omega - \frac{k^2}{2m} + \lambda\right)a^*(p)a(p) \tag{2.8}$$

$$S_1 = -\frac{1}{4\beta V}\sum_{p_1+p_2=p_3+p_4}(v(\mathbf{k}_1-\mathbf{k}_3)+v(\mathbf{k}_1-\mathbf{k}_4))a^*(p_1)a^*(p_2)a(p_4)a(p_3). \tag{2.9}$$

Here p denotes the set $(\mathbf{k},\omega)$ and $v(\mathbf{k})$ is the Fourier transform of $u(\mathbf{x})$ defined by

$$u(\mathbf{x}) = V^{-1}\sum_{\mathbf{k}} e^{i\mathbf{k}\mathbf{x}}v(\mathbf{k}). \tag{2.10}$$

The use of the Fourier coefficients $a(p)$, $a^*(p)$ gives an exceptionally simple form to the functional integral as well as to diagram perturbation theory. In terms of a, a^* we have the following explicit form for the measure:

$$\mathrm{D}\bar{\psi}\mathrm{D}\psi = \prod_p \mathrm{d}a^*(p)\,\mathrm{d}a(p). \tag{2.11}$$

The functional integral can be defined as a limit of the finite-dimensional integral over the variables $a(p)$, $a^*(p)$ with

$$|\mathbf{k}| < k_0, |\omega| < \omega_0, \quad \text{when } k_0, \omega_0 \to \infty.$$

One-particle Green's function (2.5) clearly depends on the differences $\mathbf{x}-\mathbf{x}'$, $\tau-\tau'$. Substituting the Fourier expansions (2.1) for ψ, $\bar{\psi}$ we come to the conclusion that (2.5) can be expressed in terms of the averages

$$G(p) = -\langle a(p)a^*(p)\rangle$$

$$= -\frac{\int e^S a(p)a^*(p)\prod_p \mathrm{d}a^*(p)\mathrm{d}a(p)}{\int e^S\prod_p \mathrm{d}a^*(p)\mathrm{d}a(p)}. \tag{2.12}$$

Now we can build up the perturbative scheme for Green's functions such as (2.5) and (2.12). First of all let us calculate the generating functional for the free field theory, i.e. when the pair potential $u(\mathbf{x}-\mathbf{y})$ vanishes

$$Z_0[\eta,\eta^*]=\frac{\int\prod_p da^*(p)da(p)\exp\left[S_0+\sum_p(\eta^*(p)a(p)+a^*(p)\eta(p))\right]}{\int\prod_p da^*(p)da(p)\exp S_0}. \tag{2.13}$$

This expression is equal to the product over p of the ratios of two-dimensional integrals. It is not hard to calculate each of them. The result is

$$Z_0[\eta,\eta^*]=\exp\left(-\sum_p\eta^*(p)G_0(p)\eta(p)\right) \tag{2.14}$$

where

$$G_0(p)=\left(i\omega-\frac{k^2}{2m}+\lambda\right)^{-1} \tag{2.15}$$

is nothing but the inverse of the coefficient function in the quadratic form S_0.

Differentiating (2.14) first with respect to $\eta^*(p)$, then with respect to $\eta(p)$ and putting $\eta=\eta^*=0$, we receive an expression for the unperturbed Green function:

$$\begin{aligned}G_0(p)&=-\frac{\int a(p)a^*(p)e^{S_0}\prod_p da^*(p)da(p)}{\int e^{S_0}\prod_p da^*(p)da(p)}\\&=\left(i\omega-\frac{k^2}{2m}+\lambda\right)^{-1}.\end{aligned} \tag{2.16}$$

Moreover differentiating (2.14) n times with respect to $\eta(p_i)$ and n times with respect to $\eta^*(p_j)$ $(j=1,\ldots,n)$ and then putting $\eta=\eta^*=0$, we obtain the expression for the average

$$\left\langle\prod_{i=1}^n a(p_i)\prod_{j=1}^n a^*(p_j)\right\rangle_0, \tag{2.17}$$

where

$$\langle f(a,a^*)\rangle_0=\frac{\int f(a,a^*)e^{S_0}\prod_p da^*(p)da(p)}{\int e^{S_0}\prod_p da^*(p)da(p)}. \tag{2.18}$$

From the right-hand side of (2.14) it is clear that the average (2.17) is equal to the sum of products of all possible pair averages

$$\langle a(p_i)a^*(p_j)\rangle_0=-\delta_{p_ip_j}G_0(p_i) \tag{2.19}$$

(the sum over all possible ways to select n pairs a_i, a_j^* out of n objects a_i and n objects a_j^*).

This statement is known as *Wick's theorem*. For instance, if $n=2$, we have

$$\langle a(p_1)a(p_2)a^*(p_3)a^*(p_4)\rangle_0 = \langle a(p_1)a^*(p_3)\rangle_0\langle a(p_2)a^*(p_4)\rangle_0 + \langle a(p_1)a^*(p_4)\rangle_0\langle a(p_2)a^*(p_3)\rangle_0 \tag{2.20}$$

(the averages $\langle a\ a\rangle_0$, $\langle a^*\ a^*\rangle_0$ vanish).

It is easily seen that (2.14) is equivalent to Wick's theorem.

Now we can develop the perturbation theory. The starting point is the expansion

$$e^S = e^{S_0}e^{S_1} = e^{S_0}\sum_{n=0}^{\infty}\frac{S_1^n}{n!}, \tag{2.21}$$

where S_0 is a quadratic form, S_1 is a fourth-order form in the functional S. We express Green's function (2.12) as a ratio of two series

$$G(p) = -\frac{\sum_{n=0}^{\infty}\frac{1}{n!}\langle a(p)a^*(p)S_1^n\rangle_0}{\sum_{n=0}^{\infty}\frac{1}{n!}\langle S_1^n\rangle_0}. \tag{2.22}$$

Each of the S_1 inside $\langle\cdots\rangle_0$ is a fourth-order form in the integration variables a, a^*. Thus, we deal with the expectation values

$$\left\langle\prod_{i=1}^{n} a^*(p_{1i})a^*(p_{2i})a(p_{3i})a(p_{4i})\right\rangle_0 \tag{2.23}$$

in the denominator of (2.22). The following expectation values:

$$\left\langle a(p)a^*(p)\prod_{i=1}^{n} a^*(p_{1i})a^*(p_{2i})a(p_{3i})a(p_{4i})\right\rangle_0 \tag{2.24}$$

appear in the numerator.

Here we need Wick's theorem, in order to write down the averages $\langle\cdots\rangle_0$ such as (2.23) or (2.24) as sums of products of all possible pair averages. This allows us to calculate every term in the series in the numerator and the denominator.

Feynman suggested assigning a picture diagram to each term of the series in quantum field theory analogous to (2.22). A perturbation theory in which every term corresponds to a diagram is called the diagram technique. It is not difficult to build up the diagram technique in statistical physics as well. For the case of a system of Bose particles with pair interaction we can go over to diagrams as follows. To the average (2.23) we assign a diagram made of n vertices of fourth-order (points with two incoming and two outgoing arrows). For $n = 2$ the diagram has the form

(2.25)

The incoming arrows correspond to the variable a, and the outgoing ones to a^*. To the average (2.24) we assign the diagram

(2.26)

Diagrams introduced in such a way will be called prediagrams in order to distinguish them from those which will arise later.

In accordance with Wick's theorem, the expectation values (2.22), (2.23) are sums of all possible pair averages. To every type of composing the pair average we assign a diagram by connecting each pair of the vertices i, j with a line if the average

$$\langle a(p_i)a^*(p_j)\rangle_0$$

is present among the pair expectation values. For $n = 2$ we can list all the diagrams arising from prediagram (2.25)

(2.27)

and also all diagrams arising from prediagram (2.26)

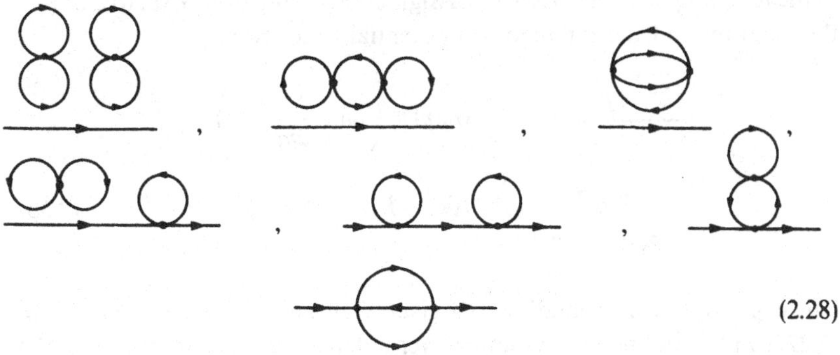

(2.28)

We get the expression corresponding to a given diagram if the product of the pair expectation values is multiplied by

$$\frac{(-1)^n}{n!}\prod_{i=1}^{n}\frac{(v(\mathbf{k}_{1i}-\mathbf{k}_{3i})+v(\mathbf{k}_{1i}-\mathbf{k}_{4i}))}{4\beta V} \tag{2.29}$$

and also by the number of ways through which a given diagram can be obtained from a prediagram. Then we have to perform the summation over all independent four-momenta p_i of internal lines.

It is convenient to reformulate slightly the resulting rules of correspondence. The coefficient of $a^*(p_1)a^*(p_2)a(p_3)a(p_4)$ in S_1 (2.9) is symmetric with respect to the permutations

$$p_1 \rightleftarrows p_2 \quad \text{and} \quad p_3 \rightleftarrows p_4. \tag{2.30}$$

The demonstrated symmetry allows us to speak about the prediagram symmetry group of the order

$$R = 4^n n! \tag{2.31}$$

The factor 4^n corresponds to symmetry groups at each fourth-order vertex; $n!$ corresponds to the group of vertex permutations. Let us note that in (2.29) we just have exactly R^{-1}. It is easily seen that

$$N = \frac{R}{r}, \tag{2.32}$$

where N is the number of ways to construct a given diagram from a prediagram, r is the order of the symmetry group of the diagram. So we obtain the factor r^{-1} in the expression corresponding to a given diagram. The presence of the symmetry factor r^{-1} is a common feature in diagram perturbation technique in quantum field theory and statistical physics.

The above arguments allow us to formulate the rules of correspondence in the following manner. We shall assign Green's function $G_0(p)$ to a line of the diagram and the symmetrized potential to a vertex

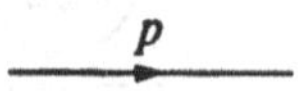

$$G_0(p) = \left(i\omega - \frac{k^2}{2m} + \lambda\right)^{-1}$$

$$v(\mathbf{k}_1-\mathbf{k}_3)+v(\mathbf{k}_1-\mathbf{k}_4). \tag{2.33}$$

The expression corresponding to a given diagram for the denominator in (2.22) can be obtained by summing over independent four-momenta of the product of expression which correspond to lines and vertices of the diagram and then multiplying the result by

$$r^{-1}\left(\frac{-1}{\beta V}\right)^{l-v}, \tag{2.34}$$

where l is the number of lines, v is the number of vertices, r is the order of the diagram symmetry group. For the diagram corresponding to the numerator the factor in front of the diagram is equal to

$$r^{-1}\left(\frac{-1}{\beta V}\right)^{l-v-1}, \tag{2.35}$$

where $l-v-1=c$ is the number of independent loops in the diagram.

It is a well-known fact that the denominator in (2.22) can be written in exponential form:

$$\sum_{n=0}^{\infty}\frac{1}{n!}\langle S_1^n\rangle_0=\exp\left(\sum_i D_i^{\mathrm{vac}}\right). \tag{2.36}$$

Here D_i^{vac} is the contribution of the ith connected vacuum diagram (i.e. a diagram without the outgoing lines). The exponential is due to the symmetry factor

$$r^{-1}=\prod_i (n_i!r_i^{n_i})^{-1} \tag{2.37}$$

for any diagram which consists of n_1 connected components of the first kind, n_2 components of the second kind and so on. The numerator of (2.22) is equal to the factor (2.36) multiplied by the sum of all connected diagrams without vacuum components. This is why the factor (2.36) in the numerator cancels that in the denominator, and so we can take into account only connected diagrams without vacuum components. The transition to connected diagrams is a common fact as well as the above-mentioned appearance of the factor r^{-1}.

The diagram techniques developed above in the functional integral approach coincide with the well-known Matsubara–Abrikosov–Gorkov–Dzyaloshinsky perturbation theory for the temperature Green functions. The functional integral method gives us the shortest way to obtain it. Moreover, functional methods are very useful for a reformulation of perturbation theory in many cases when the perturbative scheme is not applicable in its standard form, for instance for superfluid Bose or Fermi systems. From the functional integration point of view such a reformulation is an alternative method of asymptotic evaluation of the functional integral.

Now let us dwell on some modifications of the diagram techniques connected with different methods of partial summation of diagrams. As we have seen, the vacuum diagrams give no contribution to the expressions for the Green functions so that when dealing with Green functions we may only take into account connected diagrams.

It is well-known that the Green function $G(p)$ can be expressed in terms of its irreducible self energy part $\Sigma(p)$ ($\Sigma(p)$ is the sum of all diagrams with two tails

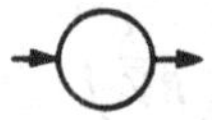

such that it is impossible to cut the diagram into two without cutting one line of the diagram). So we can write down the following equation

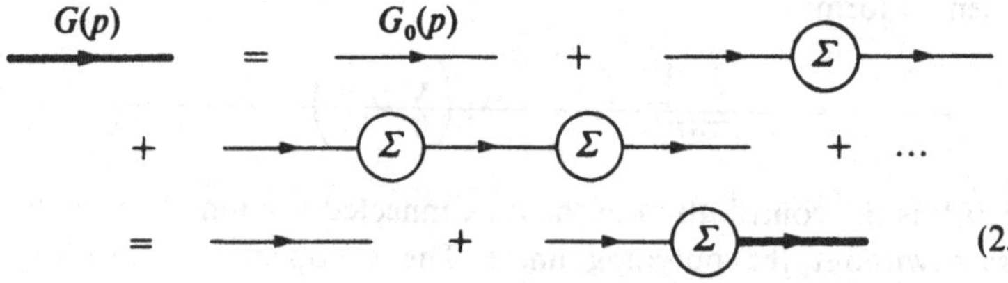

(2.38)

This coincides with the well-known Dyson equation. Its analytic form is as follows:

$$G(p) = G_0(p) + G_0(p)\Sigma(p)G(p). \tag{2.39}$$

Its solution is

$$G(p) = (G_0^{-1}(p) - \Sigma(p))^{-1}. \tag{2.40}$$

Thus in order to determine the full Green function it is sufficient to find its irreducible self energy part.

There is a well-known method of summation of diagrams which is connected with the Dyson formula. This is the so-called skeleton diagram techniques. We come to this form of perturbative scheme if we perform partial summation of diagrams which is equivalent to replacing the bare Green function $G_0(p)$ by the full Green functions $G(p)$ in all inner lines of the diagrams. So the elements of the skeleton diagram techniques are the full Green functions and bare vertices

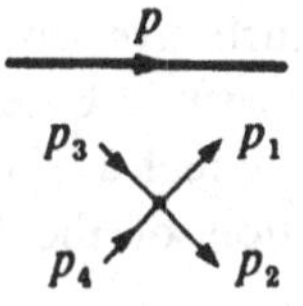

$G(p)$

$$v(\mathbf{k}_1 - \mathbf{k}_3) + v(\mathbf{k}_1 - \mathbf{k}_4). \tag{2.41}$$

Here we need not take into account diagrams with self-energy part insertions into inner lines. Of course, we do not know $G(p)$ from the very beginning, so we have to deal with a system of equations which would allow us to find $G(p)$. The first of them is the Dyson equation (2.40), and the second one is the equation for the irreducible self-energy part

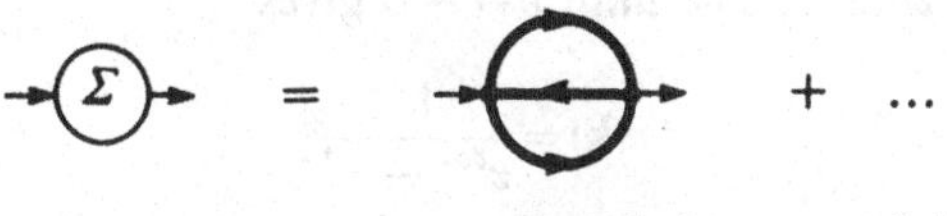

(2.42)

Here on the right-hand side we have the infinite sum of diagrams of the skeleton diagram techniques. Such modified diagram techniques are suitable, for instance, in theories with the anomalous Green functions, which are identically equal to zero if calculated by the standard scheme of perturbation theory. Such a situation takes place in the theory of superfluidity and superconductivity, where there exist anomalous Green functions below the phase transition temperature.

To conclude this section we shall show how an information about physical properties of the system can be obtained by using the Green functions.

Let us consider the Bose system. We can obtain the average number of particles N in the system by averaging the functional

$$\int \bar{\psi}(\mathbf{x}, \tau)\psi(\mathbf{x}, \tau)\mathrm{d}^3\mathbf{x}. \tag{2.43}$$

This implies the formula

$$N = \frac{\int e^S[\int \bar{\psi}(\mathbf{x}, \tau)\psi(\mathbf{x}, \tau)\mathrm{d}^3x]D\bar{\psi}D\psi}{\int e^S D\bar{\psi}D\psi} = \int \mathrm{d}^3x\langle \psi(\mathbf{x}, \tau)\bar{\psi}(\mathbf{x}, \tau)\rangle. \tag{2.44}$$

If our system is a translation-invariant one, the Green function (2.6) can only depend on $\mathbf{x} - \mathbf{x}', \tau - \tau'$. So we come to the following result:

$$N = V \lim_{\substack{\mathbf{x}' \to \mathbf{x} \\ \tau' \to \tau}} \langle \psi(\mathbf{x}, \tau)\bar{\psi}(\mathbf{x}', \tau')\rangle. \tag{2.45}$$

It turns out that the result depends on the way of taking the limit. Let us demonstrate this for the ideal Bose system. Here we have

$$\begin{aligned}\langle \psi(\mathbf{x}, \tau)\bar{\psi}(\mathbf{x}', \tau')\rangle &= (\beta V)^{-1} \sum_{\mathbf{k},\omega} G_0(p) \exp \mathrm{i}(\omega(\tau - \tau') + \mathbf{k}(\mathbf{x} - \mathbf{x}')) \\ &= (\beta V)^{-1} \sum_{\mathbf{k},\omega} \frac{\exp \mathrm{i}(\omega(\tau - \tau') + \mathbf{k}(\mathbf{x} - \mathbf{x}'))}{i\omega - \dfrac{k^2}{2m} + \lambda}.\end{aligned} \tag{2.46}$$

Putting here $\mathbf{x}' = \mathbf{x}, \tau' = \tau - \varepsilon\ (\varepsilon > 0)$, and summing over the frequencies ω we find

$$\beta^{-1}\sum_{\omega}\frac{e^{i\omega\varepsilon}}{i\omega - \varepsilon(\mathbf{k})} = \frac{e^{\varepsilon\varepsilon(\mathbf{k})}}{e^{\beta\varepsilon(\mathbf{k})} - 1}, \tag{2.47}$$

where $\varepsilon(\mathbf{k}) = k^2/2m - \lambda$. The limit $\varepsilon \to +0$ gives

$$n(\mathbf{k}) = \frac{1}{e^{\beta\varepsilon(\mathbf{k})} - 1}, \tag{2.48}$$

which leads to the well-known formula of the Bose-partition

$$N = \sum_{\mathbf{k}} n(\mathbf{k}) = \sum_{\mathbf{k}} \frac{1}{e^{\beta\varepsilon(\mathbf{k})} - 1}. \tag{2.49}$$

If we take the limit $\varepsilon \to -0$, we come to a different result:

$$\frac{1}{e^{\beta\varepsilon(\mathbf{k})} - 1} + 1 \tag{2.50}$$

instead of (2.48). The correct answer is given by the first limit. This is also correct for a nonideal system and we get the following formula for the density of particles:

$$\rho = \frac{N}{V} = -\lim_{\varepsilon \to +0}(\beta V)^{-1}\sum_{p} e^{i\omega\varepsilon}G(p). \tag{2.51}$$

The momentum and kinetic energy of the system can be obtained by averaging the following functionals:

$$\frac{i}{2}\int(\nabla\bar{\psi}\psi - \bar{\psi}\nabla\psi)\,d^3x \qquad \text{momentum} \tag{2.52}$$

$$\frac{1}{2m}\int \nabla\bar{\psi}\nabla\psi\,d^3x \qquad \text{kinetic energy.} \tag{2.53}$$

This leads to the following formulae:

$$\frac{\mathbf{K}}{V} = -(\beta V)^{-1}\sum_{p} e^{i\omega\varepsilon}\mathbf{k}G(p) \tag{2.54}$$

$$\frac{H_{\mathrm{kin}}}{V} = -(\beta V)^{-1}\sum_{p} e^{i\omega\varepsilon}\frac{k^2}{2m}G(p). \tag{2.55}$$

We have to use the prescription $\varepsilon \to +0$ in both (2.54) and (2.55).

We shall derive further the expression for pressure, starting from a formula for the ratio of partition sums of nonideal and ideal systems:

$$\frac{\int e^{S} D\bar{\psi} D\psi}{\int e^{S_0} D\bar{\psi} D\psi} = \frac{Z}{Z_0} = \exp \beta(\Omega_0 - \Omega). \tag{2.56}$$

Here Z_0, Z are partition functions of ideal and nonideal systems, $\Omega_0 = -p_0 V, \Omega = -pV$, where p_0 is the pressure of the ideal system, p is the pressure of the nonideal one. The left-hand side of (2.56) is equal to

$$\exp\left(\sum_i D_i^{\text{vac}}\right), \tag{2.57}$$

where $\sum_i D_i^{(\text{vac})}$ is the sum of all connected vacuum diagrams. We thus have

$$p = p_0 + (\beta V)^{-1} \sum_i D_i^{\text{vac}}. \tag{2.58}$$

This formula expresses the pressure p in terms of p_0 and the sum of contributions of all vacuum diagrams.

3

Functional integrals and diagram techniques for Fermi particles

In the previous section we discussed a quantization scheme for Bose fields in the functional integral approach. Quantization of Fermi fields may be performed using the functional integral over anticommuting variables (for details see Berezin, 1966).

We can define the integral over Fermi fields as a limit of the integral over the Grassmann algebra with a finite (even) number of generators

$$x_i, x_i^* \quad i = 1, \ldots, n, \tag{3.1}$$

which anticommute with each other

$$\begin{aligned} x_i x_j + x_j x_i &= 0, \\ x_i^* x_j^* + x_j^* x_i^* &= 0, \\ x_i x_j^* + x_j^* x_i &= 0. \end{aligned} \tag{3.2}$$

According to (3.2) we have $x_i^2 = 0$, $(x_i^*)^2 = 0$, and each element of the algebra can be written in the following form:

$$f(x, x^*) = \sum_{a_i, b_i = 0, 1} c_{a_1 \ldots a_n b_1 \ldots b_n} x_1^{a_1} \cdots x_n^{a_n} (x_n^*)^{b_n} \cdots (x_1^*)^{b_1}. \tag{3.3}$$

Here $c_{a_1 \ldots a_n b_1 \ldots b_n}$ are complex coefficients.

Let us define the involution operation in the algebra by the formula:

$$f \to f^* = \sum_{a_i, b_i = 0, 1} \bar{c}_{a_1 \ldots a_n b_1 \ldots b_n} x_1^{b_1} \cdots x_n^{b_n} (x_n^*)^{a_n} \cdots (x_1^*)^{a_1}. \tag{3.4}$$

We can now introduce the functional integral over the algebra

$$\int f(x, x^*) \mathrm{d}x^* \mathrm{d}x = \int f(x_1, \ldots, x_n, x_1^*, \ldots, x_n^*) \mathrm{d}x_1^* \mathrm{d}x_1 \cdots \mathrm{d}x_n^* \mathrm{d}x_n. \tag{3.5}$$

The definition is made precise by setting

$$\int \mathrm{d}x_i = 0, \quad \int \mathrm{d}x_i^* = 0, \quad \int x_i \mathrm{d}x_i = 1, \quad \int x_i^* \mathrm{d}x_i^* = 1 \tag{3.6}$$

$$\int (c_1 f_1 + c_2 f_2) \mathrm{d}x^* \mathrm{d}x = c_1 \int f_1 \mathrm{d}x^* \mathrm{d}x + c_2 \int f_2 \mathrm{d}x^* \mathrm{d}x. \tag{3.7}$$

The symbols $\mathrm{d}x_i$, $\mathrm{d}x_i^*$ must anticommute with each other and with the generators of the algebra. The coefficients c_1, c_2 in (3.7) are complex numbers.

So integrating the polynomial function (3.3), we obtain

$$\int f \mathrm{d}x^* \mathrm{d}x = c_{1\cdots 1,1\cdots 1}. \tag{3.8}$$

The following two formulae will be important for future applications:

$$\int \exp(-x^*Ax)\mathrm{d}x^*\mathrm{d}x = \det A, \tag{3.9}$$

$$\frac{\int \exp(-x^*Ax + \eta^*x + x^*\eta)\mathrm{d}x^*\mathrm{d}x}{\int \exp(-x^*Ax)\mathrm{d}x^*\mathrm{d}x} = \exp(\eta^*A^{-1}\eta). \tag{3.10}$$

Here x^*Ax is a quadratic form,

$$x^*Ax = \sum_{i,k} a_{ik}x_i^*x_k, \tag{3.11}$$

and η^*x, $x^*\eta$ are linear forms

$$\eta^*x = \sum_i \eta_i^*x_i, \quad x^*\eta = \sum_i x_i^*\eta_i. \tag{3.12}$$

The symbols η_i, η_i^* anticommute with each other and with the generators x_i, x_i^*. We may thus regard the set $\eta_i, \eta_i^*, x_i, x_i^*$ as a set of generators of a larger algebra. The expression $\eta^*A^{-1}\eta$ in (3.10) is the quadratic form of the matrix A^{-1} inverse to A.

The exponentials in the integrands of (3.9) and (3.10) can be expressed through the expansion into series, in which, due to commutation relations (3.2), only several first terms are nonzero.

We can prove (3.9) by expanding the exponential function and then noticing that only the nth term gives a contribution to the integral. As for (3.10), we can prove it by using the shift transformation $x \to x + \tilde{\eta}, x^* \to x^* + \tilde{\eta}^*$, which cancels the linear terms in the exponent of the integrand.

Now we shall briefly discuss the functional integral and diagram technique for Fermi systems. The quantization of a Fermi system can be obtained as a result of integration over the space of anticommuting functions $\psi(\mathbf{x}, \tau)$ (the elements of an infinite Grassmann algebra), where $\mathbf{x} \in V, \tau \in [0, \beta]$. To obtain the correct result it is necessary to impose on $\psi, \bar{\psi}$ the antiperiodicity conditions in the variable τ

$$\psi(\mathbf{x}, \beta) = -\psi(\mathbf{x}, 0), \quad \bar{\psi}(\mathbf{x}, \beta) = -\bar{\psi}(\mathbf{x}, 0). \tag{3.13}$$

Then we have the following Fourier series for ψ, $\bar{\psi}$ in the Fermi case:

$$\psi(\mathbf{x},\tau) = (\beta V)^{-1/2} \sum_p e^{i(\omega\tau + \mathbf{kx})} a(p),$$

$$\bar{\psi}(\mathbf{x},\tau) = (\beta V)^{-1/2} \sum_p e^{-i(\omega\tau + \mathbf{kx})} a^*(p), \tag{3.14}$$

where $\omega = (2n+1)\pi/\beta$ are the 'Fermi frequencies'. Let us notice that the Fourier coefficients $a(p)$, $a^*(p)$ in (3.14) may be considered as generators of an infinitely dimensional Grassmann algebra.

Green's functions for Fermi systems are defined by the same formulae (2.6), (2.12) as for Bose systems, namely as the ratio of two functional integrals. Such a ratio can be understood as a limit of the ratio of two finite-dimensional functional integrals arising when only the coefficients with $|\mathbf{k}| < k_0$ and $|\omega| < \omega_0$ are taken. These integrals coincide with the previously defined integrals over the Grassmann algebra with generators $a(p)$, $a^*(p)$. It then remains to take the limit as $k_0, \omega_0 \to \infty$.

The S-functional and the Hamiltonian have the same form as those for the Bose system. For instance, the unperturbed action S_0 has the form

$$S_0 = \sum_{\mathbf{k},\omega} \left(i\omega - \frac{k^2}{2m} + \lambda \right) a^*(p) a(p), \tag{3.15}$$

where $\omega = (2n+1)\pi/\beta$ is the Fermi frequency. It is convenient to write the perturbative term S_1 as

$$S_1 = -\frac{1}{4\beta V} \sum_{p_1+p_2=p_3+p_4} (v(\mathbf{k}_1 - \mathbf{k}_3) - v(\mathbf{k}_1 - \mathbf{k}_4)) a^*(p_1) a^*(p_2) a(p_4) a(p_3), \tag{3.16}$$

with the antisymmetrized potential $v(\mathbf{k}_1 - \mathbf{k}_3) - v(\mathbf{k}_1 - \mathbf{k}_4)$.

The derivation of the diagram technique is completely analogous to that performed above for Bose systems. We come to the diagram technique with the elements

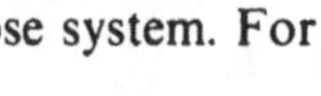

$$\left(i\omega - \frac{k^2}{2m} + \lambda \right)^{-1} \qquad \omega = (2n+1)\pi/\beta$$

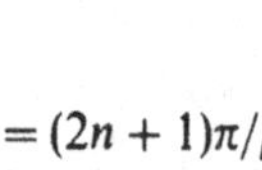

$$v(\mathbf{k}_1 - \mathbf{k}_3) - v(\mathbf{k}_1 - \mathbf{k}_4). \tag{3.17}$$

The 'Fermi' diagram techniques differ from the Bose case in the following points:

(1) The Fermi frequencies $\omega = (2n+1)\pi/\beta$ are odd, whereas the Bose frequencies are even, $\omega = 2\pi n/\beta$.

(2) We have an antisymmetrized potential $v(\mathbf{k}_1 - \mathbf{k}_3) - v(\mathbf{k}_1 - \mathbf{k}_4)$ in (3.17) instead of a symmetrized one in the Bose case.

(3) The sum over independent momenta is multiplied by the factor

$$(-1)^F r^{-1} \left(\frac{-1}{\beta V} \right)^{l-v} \tag{3.18}$$

for vacuum diagrams and by

$$(-1)^F r^{-1} \left(\frac{-1}{\beta V} \right)^{l-v-1} \tag{3.19}$$

for diagrams corresponding to the one-particle Green function. Expressions (3.18) and (3.19) differ from (2.34) and (2.35) for the Bose system by the factor $(-1)^F$, where F is the number of closed Fermi loops of the diagram. The presence of the factor $(-1)^F$ is a consequence of the fact that the fields ψ, $\bar{\psi}$ anticommute.

The formula for the pressure (2.58) is valid for both Bose and Fermi systems. The formulae for the mean number of particles, momentum and kinetic energy per unit volume for the Fermi system differ from the corresponding formulae for the Bose system (2.51), (2.54) and (2.55), replacing the Bose frequencies $\omega = 2\pi n/\beta$ with the Fermi ones $\omega = (2n+1)\pi/\beta$ and have opposite signs.

4

Method of successive integration over fast and slow variables

A lot of problems of quantum field theory and statistical physics are connected with the evaluation of the Green functions at small energies and momenta (the infrared asymptotics). If quanta with arbitrary small energies are present in the system, the standard perturbation theory leads to difficulties. Such a situation takes place in quantum electrodynamics and in various branches of statistical physics – superfluidity, superconductivity, plasma theory, the theory of ^{3}He, and the general theory of phase transitions. The difficulties mentioned stem from the fact that every graph of the ordinary perturbation theory has a so-called infrared singularity. It means that the expression assigned to a graph that is an integral over momentum variables is singular when the external momenta tend to zero. In such cases a modification of the perturbation theory is desirable. One possible modification may be called the method of subsequent integration, first over 'fast' and then over 'slow' variables.

We shall represent every field ψ that is integrated over as a sum of two terms. One of them will be called the slowly oscillating field ψ_0, and the other the fast oscillating field ψ_1

$$\psi = \psi_0 + \psi_1. \tag{4.1}$$

In the Bose-systems theory the function $\psi = \psi(\mathbf{x}, \tau)$ is supposed to be periodic in its arguments and it is decomposed into the Fourier series

$$\psi(\mathbf{x}, \tau) = (\beta V)^{-1/2} \sum_{\mathbf{k},\omega} e^{i(\omega\tau + \mathbf{kx})} a(\mathbf{k}, \omega). \tag{4.2}$$

The sum of terms in (4.2) with

$$|\mathbf{k}| \leqslant k_0 \quad \text{and} \quad |\omega| \leqslant \omega_0 \tag{4.3}$$

will be called the slowly oscillating part $\psi_0(\mathbf{x}, \tau)$ of $\psi(\mathbf{x}, \tau)$. The difference $\psi(\mathbf{x}, \tau) - \psi_0(\mathbf{x}, \tau)$ will be called the fast oscillating part $\psi_1(\mathbf{x}, \tau)$. This difference is, of course, the sum of terms in (4.2) with $|\mathbf{k}| > k_0$ or $|\omega| > \omega_0$.

It is worth noting that the boundary between the slow and fast fields in the concrete problems of quantum field theory and statistical physics is to some extent conditional. It reflects in the fact that the parameters k_0, ω_0 distinguishing between the 'slow' and 'fast' variables are not determined exactly but only in order.

In the theory of Fermi systems the decomposition of Fermi fields has the same form (4.2), as that for Bose systems, but with Fermi frequencies $\omega = (2n+1)\pi/\beta$ replacing the Bose ones $\omega = 2\pi n/\beta$. In many cases it is natural to call the sum of terms in (4.2) with

$$|k - k_F| < k_0, \quad |\omega| < \omega_0 \tag{4.4}$$

the 'slow' part of the Fermi field $\psi(\mathbf{x}, \tau)$. Here the momenta $\mathbf{k}$ belong to the narrow shell around the Fermi sphere $k = k_F$. Such a definition of 'slow' and 'fast' variables turns out to be very useful in the theory of superfluid Fermi systems.

As we know, the functional integral in statistical physics is the integral with the measure

$$\prod_{\mathbf{k},\omega} da^*(\mathbf{k},\omega)da(\mathbf{k},\omega). \tag{4.5}$$

The fundamental idea of the modification of the perturbation theory consists in the successive integration, at first over the 'fast' field and afterwards over the 'slow' one, using different schemes of perturbation theory at the two different stages of integration. At the first stage we integrate over the Fourier coefficients $a(\mathbf{k},\omega)$, $a^*(\mathbf{k},\omega)$, whose indices $(\mathbf{k},\omega)$ lie in the region of fast variables. At the second stage integration over slow variables is carried out.

When integrating over the fast fields we can use the perturbation theory and corresponding graphs that differ from the standard one explained in the previous sections in two points:

(1) The integrals (sums) over the four-momenta are cut off at a low limit.

(2) Supplementary vertices will emerge describing the interaction of the fast field ψ_1 with the slow field ψ_0.

The first point is obvious because the variables (k, ω) of the fast field lie in the 'fast' region. The supplementary vertices appear because the starting action S is expressed via $\psi = \psi_0 + \psi_1$ so that the crossing terms that may not appear in the quadratic form do contribute to the terms of the third and higher degrees.

Cutting off integrals at a low limit prevents the emergence of infrared divergencies at the first stage of the integration. At the second stage (integration over the slow fields) we may achieve the vanishing of infrared divergencies if a nontrivial perturbative scheme exploiting the special features of the system is adopted. In the superfluidity theory, for example, when integrating over the slow Bose fields $\psi_0(\mathbf{x},\tau)$, $\bar{\psi}_0(\mathbf{x},\tau)$ it proves convenient to pass to the polar coordinates

$$\psi_0(\mathbf{x},\tau)=(\rho(\mathbf{x},\tau))^{1/2}e^{i\varphi(\mathbf{x},\tau)},\quad \bar{\psi}_0(\mathbf{x},\tau)=(\rho(\mathbf{x},\tau))^{1/2}e^{-i\varphi(\mathbf{x},\tau)} \tag{4.6}$$

and integrate over the field $\rho(\mathbf{x},\tau)$, $\varphi(\mathbf{x},\tau)$. The perturbation theory for the integral over slow fields will be formulated in terms of Green's function of the fields $\rho(\mathbf{x},\tau)$, $\varphi(\mathbf{x},\tau)$. As will be shown in the following sections, such a perturbative scheme is free from infrared divergencies.

In the case of superfluid Fermi systems it turns out to be convenient to pass from the integral over slow Fermi fields to the integral over some auxiliary Bose fields corresponding to the 'collective' degrees of freedom of the system considered.

In what follows we give concrete illustrations of how one can implement the idea of the successive integration first over fast and then over slow variables. This approach proves to be very effective for the description of collective excitations in both Bose and Fermi superfluid systems and also for many other physically interesting cases.

Part II

Superfluid Bose systems

5
Superfluidity

Here and in the three sections that follow we will consider some applications of functional integration methods to the theory of interacting Bose particles. Such systems may undergo a phase transition into the superfluid state in which long-range correlations arise. This means that the average

$$\langle \psi(\mathbf{x}, \tau)\bar{\psi}(\mathbf{x}', \tau') \rangle \tag{5.1}$$

does not decrease exponentially as $r = |\mathbf{x} - \mathbf{x}'| \to \infty$.

In three-dimensional systems the average (5.1) has a positive limit ρ_0 (the so-called density of the condensate) as $r \to \infty$. In two-dimensional systems at low temperatures and in one-dimensional ones at $T = 0$ (5.1) decreases as $r^{-\alpha}$. The exponent α can be expressed in terms of the thermodynamical functions of the system in question. Two-dimensional and one-dimensional cases will be considered in sections 7 and 8.

We begin with three-dimensional systems. The problem of a nonideal Bose gas at $T = 0$ was solved by N. N. Bogoliubov (Bogoliubov, 1947). Later he and his collaborators D. N. Zubarev, Yu. A. Tserkovnikov and V. V. Tolmachev considered the higher approximations and the case of $T \neq 0$ (Bogoliubov & Zubarev, 1955; Tolmachev, 1960*a–c*, 1969; Tserkovnikov, 1962, 1964*a, b*). Quantum field theory methods were applied to the theory of low-density Bose gas by S. T. Beliaev (Beliaev, 1958*a, b*) and also by N. M. Hugenholtz and D. Pines (Hugenholtz & Pines, 1959). The Bose gas problem at $T \neq 0$ was studied by T. D. Lee and C. N. Yang (Lee & Yang, 1958, 1959*a–c*, 1960*a, b*) who used the so-called pseudopotential method.

First we will consider the simplest variant of perturbation theory which takes into account the possibility of Bose condensation in the system. This perturbative scheme is nothing but a slight modification of that developed in the framework of the functional integration method in section 2. In the operator formalism this variant of the diagram technique was suggested by Popov & Faddeev (1964) (see also Popov, 1964, 1965).

This perturbation theory suffices for many purposes. Still it is not a perfect one, because its diagrams are singular at low momenta. Summation of such diagrams in regions of low momentum is a fairly difficult problem.

Here we will consider the simplest perturbative scheme and its applications to the theory of low-density Bose gas. It will be shown why the diagrams of the theory are singular at low momenta. In section 7 a new perturbation method is developed which permits the elimination of low-momentum (infrared) singularities.

The simplest way to take Bose condensation into account is to perform the following shift transformation:

$$\psi(\mathbf{x},\tau)\to\psi(\mathbf{x},\tau)+\alpha,\quad \bar{\psi}(\mathbf{x},\tau)\to\bar{\psi}(\mathbf{x},\tau)+\bar{\alpha} \tag{5.2}$$

in the functional integral over $\psi, \bar{\psi}$ fields. Here α is a complex number. We will see that the most natural choice of α is when $|\alpha|^2$ equals the condensate density $\rho_0(|\alpha|^2=\rho_0)$. Using Fourier coefficients, we can rewrite the shift transformation (5.2) as

$$a(p)=b(p)+\alpha(\beta V)^{1/2}\delta_{p0},\quad a^*(p)=b^*(p)+\bar{\alpha}(\beta V)^{1/2}\delta_{p0}. \tag{5.3}$$

Here $\delta_{p0}=1$ if $p=0$ and $\delta_{p0}=0$ if $p\neq 0$.

Substituting (5.3) into formulae (2.8) + (2.9), we find the following S-functional:

$$\begin{aligned}S=&\beta V C_0-(\beta V)^{1/2}(\bar{\gamma}b(0)+\gamma b^*(0))+\sum_p\left(i\omega-\frac{k^2}{2m}+\lambda\right)b^*(p)b(p)\\&-\sum_p|\alpha|^2(v(0)+v(\mathbf{k}))b^*(p)b(p)-\frac{1}{2}\sum_p v(\mathbf{k})(\alpha^2b^*(p)b^*(-p)+\bar{\alpha}^2b(p)b(-p))\\&-\frac{1}{2(\beta V)^{1/2}}\sum_{p_1+p_2=p_3}(v(\mathbf{k}_1)+v(\mathbf{k}_2))(\alpha b^*(p_1)b^*(p_2)b(p_3)\\&+\bar{\alpha}b^*(p_3)b(p_2)b(p_1))\\&-\frac{1}{4\beta V}\sum_{p_1+p_2=p_3+p_4}(v(\mathbf{k}_1-\mathbf{k}_3)+v(\mathbf{k}_1-\mathbf{k}_4))b^*(p_1)b^*(p_2)b(p_3)b(p_4).\end{aligned} \tag{5.4}$$

Here

$$C_0=\lambda|\alpha|^2-\tfrac{1}{2}v(0)|\alpha|^4,\quad \gamma=\alpha(-\lambda+v(0)|\alpha|^2). \tag{5.5}$$

Let us take the quadratic form

$$\sum_p\left(i\omega-\frac{k^2}{2m}+\lambda\right)b^*(p)b(p) \tag{5.6}$$

as a free field theory form, as we did in section 2. The other forms of the first, second, third and fourth degree will be regarded as perturbations. Then we come down to the following diagram techniques:

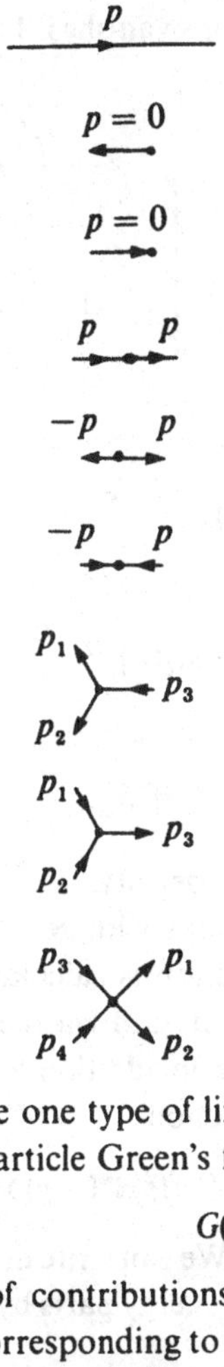

$$\left(i\omega - \frac{k^2}{2m} + \lambda\right)^{-1}$$

$$\gamma$$

$$\bar{\gamma}$$

$$|\alpha|^2(v(0) + v(\mathbf{k}))$$

$$\alpha^2 v(\mathbf{k})$$

$$\bar{\alpha}^2 v(\mathbf{k})$$

$$\alpha(v(\mathbf{k}_1) + v(\mathbf{k}_2))$$

$$\bar{\alpha}(v(\mathbf{k}_1) + v(\mathbf{k}_2))$$

$$v(\mathbf{k}_1 - \mathbf{k}_3) + v(\mathbf{k}_1 - \mathbf{k}_4). \tag{5.7}$$

Here we have one type of line and eight types of vertices.

The one-particle Green's function

$$G(p) = -\langle b(p)b^*(p)\rangle \tag{5.8}$$

is the sum of contributions of all connected diagrams. To obtain the expression corresponding to a given diagram, one must take the product of expressions corresponding to its elements, sum these products over all independent four-momenta and multiply the result by

$$\frac{1}{r}\left(\frac{-1}{\beta V}\right)^{l-v-1}, \tag{5.9}$$

where l is the number of lines, v is the number of vertices, r is the order of the diagram symmetry group and $l - v - 1 = c$ is the number of independent loops.

Let us now consider diagrams which contain parts connected to the rest of the diagram by only one line. The four-momentum of this line vanishes ($p = 0$). We shall call those parts zero-momentum insertions. We can choose α in such a way that the sum of all these insertions vanishes. Then we come to the following diagram equation:

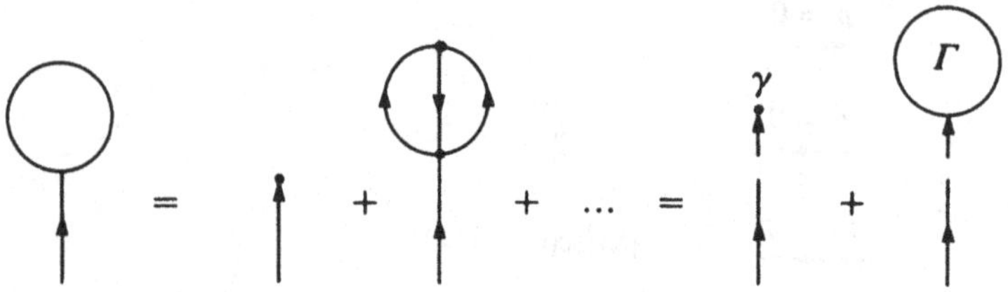

(5.10)

This is equivalent to

$$\langle b(0) \rangle = \langle b^*(0) \rangle = 0. \tag{5.11}$$

Thus we have

$$\langle a(0) \rangle = \alpha(\beta V)^{1/2}, \quad \langle a^*(0) \rangle = \bar{\alpha}(\beta V)^{1/2} \tag{5.12}$$

and

$$\langle a(p)a^*(p) \rangle = \langle b(p)b^*(p) \rangle + \beta V |\alpha|^2 \delta_{p0}. \tag{5.13}$$

We see that $|\alpha|^2$ is nothing but the condensate density.

If condition (5.10) is satisfied, then all diagrams with zero-momentum insertions can be left out when a Green's function is calculated. As was shown in section 2, Green's function can be expressed via the irreducible self-energy part. For superfluid systems we have, in addition to the normal Green's functions, two anomalous Green's functions

$$G_1(p) = -\langle a(p)a(-p) \rangle, \quad \bar{G}_1(p) = -\langle a^*(p)a^*(-p) \rangle \tag{5.14}$$

and two anomalous self-energy parts $B(p)$, $\bar{B}(p)$. We can write down G, G_1 in terms of the normal, A and anomalous $B, \bar{B}$, self-energy parts by solving the system of Dyson–Beliaev equations

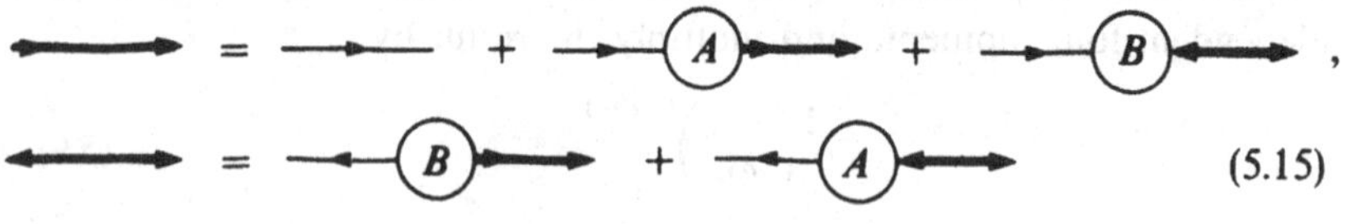

(5.15)

or in analytical form

$$G(p) = G_0(p) + G_0(p)A(p)G(p) + G_0(p)B(p)G_1(p),$$

$$G_1(p) = G_0(-p)\bar{B}(p)G(p) + G_0(-p)A(-p)G_1(p). \tag{5.16}$$

This linear system can be easily solved:

$$G(p) = \left[i\omega + \frac{k^2}{2m} - \lambda + A(-p) \right] Z^{-1}(p), \quad G_1(p) = -\bar{B}(p) Z^{-1}(p), \quad (5.17)$$

where

$$Z(p) = \left(i\omega + \frac{k^2}{2m} - \lambda + A(-p) \right)\left(i\omega - \frac{k^2}{2m} + \lambda - A(p) \right) + |B(p)|^2. \quad (5.18)$$

Now let us come back to (5.10) which allows us to determine α. Equation (5.10) can be written as

$$G_0(0)(\gamma + \Gamma) = 0, \quad (5.19)$$

which implies $\gamma + \Gamma = 0$, since $G_0(0) \neq 0$. For $\gamma + \Gamma$ we have the following exact identity (Hugenholtz–Pines formula)

$$\gamma + \Gamma = \alpha(A(0) - B(0) - \lambda). \quad (5.20)$$

This formula holds for α real and will be justified below. So we come to the equation

$$\alpha(A(0) - B(0) - \lambda) = 0 \quad (5.21)$$

which replaces (5.10). This equation has a trivial solution $\alpha = 0$ but may also have a nontrivial one if the second factor in (5.21) vanishes. For the case of a low-density Bose gas we will show that such a nontrivial solution does exist for low temperatures. Moreover, if it exists, we must deal with this solution, because the trivial one leads to discrepancies. For large temperatures we have no nontrivial solution and thus come back to the standard perturbation theory.

Let us now prove the Hugenholtz–Pines formula. First of all we notice that the differentiation of an arbitrary diagram element with respect to $\alpha(\bar{\alpha})$ would be equivalent to attaching to this element the incoming (outgoing) arrow (with $p = 0$). For instance, we have

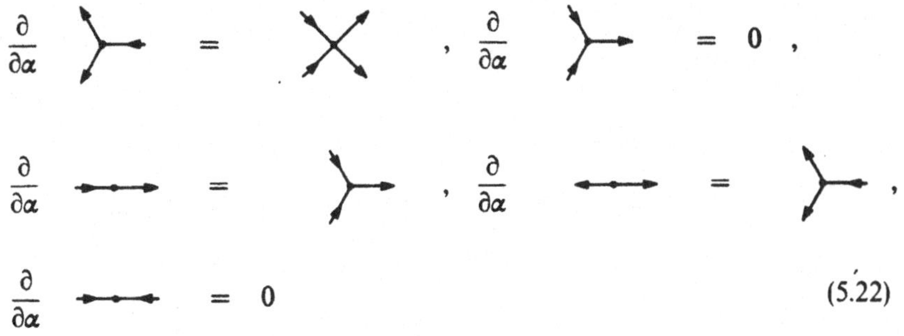

(5.22)

So the differentiation of an arbitrary diagram (with no zero-momentum insertions) with respect to $\alpha(\bar{\alpha})$ results in attaching the incoming (outgoing) arrow to the diagram in all possible ways. We can now analyse all the diagrams for $p = -\Omega/V$. We have

$$p = p_0 + C_0 + (\beta V)^{-1} \sum_i D_i^{\text{vac}}, \tag{5.23}$$

where p_0 is the pressure of an ideal system, C_0 is defined by (5.5), and $\sum_i D_i^{\text{vac}}$ is a sum of all connected vacuum diagrams. Then the following relations:

$$\gamma + \Gamma = -\frac{\partial p}{\partial \alpha}, \quad A(0) - \lambda = -\frac{\partial^2 p}{\partial \alpha \partial \bar{\alpha}},$$

$$B(0) = -\frac{\partial^2 p}{\partial \alpha^2} \tag{5.24}$$

can be proved by comparing diagrams for both sides of these equations. It is evident that p depends on α, $\bar{\alpha}$ via the product $\alpha\bar{\alpha} = z$. So we can rewrite (5.24) in the form

$$\gamma + \Gamma = -\bar{\alpha}\frac{\partial p}{\partial z}, \quad A(0) - \lambda = -\frac{\partial p}{\partial z} - z\frac{\partial^2 p}{\partial z^2},$$

$$B(0) = -\bar{\alpha}^2\frac{\partial^2 p}{\partial z^2}. \tag{5.25}$$

This implies (5.20), if α is real. Let us notice that the existence of a nontrivial solution of (5.21) amounts to the existence of an extremum for the function $p(\rho_0)$:

$$\frac{\partial p}{\partial \rho_0} = 0. \tag{5.26}$$

6
Low-density Bose gas

Here we shall be concerned with a concrete Bose system of low density, where we can calculate approximately the Green's functions. The low-density condition means that the effective potential radius a is much smaller than the mean particle distance, i.e.

$$\theta = \rho^{1/3} a \ll 1. \tag{6.1}$$

Besides (6.1), we will adopt the condition

$$T \sim \rho^{2/3}, \tag{6.2}$$

(in this section we use the systems of units $\hbar = k_B = 2m = 1$), which defines the temperature region to be considered. The point is that in the region (6.2) a phase transition from normal to superfluid state occurs.

For a low-density system we can determine the diagrams which give the main contribution into Green's functions.

It turns out that the contribution of a given diagram is diminished by the following factors:

(1) Anomalous vertices which are proportional to ρ_0 or $\rho_0^{1/2}$ (since $\rho_0 < \rho \ll a^{-3}$).

(2) Closed cycles of normal lines. We can go around such a cycle by moving along the arrows. The point is that factors $(e^{\beta\varepsilon(\mathbf{k})} - 1)^{-1}$ with $\varepsilon(\mathbf{k}) = k^2 - \lambda$ arise after summation over frequencies. These factors cut off the integrals over the k-variable effectively for $k \sim \beta^{-1/2} = T^{1/2}$, which is of the order $\rho^{1/3}$ due to (6.2) and much smaller than a^{-1} due to (6.1). So the cut-off $T^{1/2}$ is much smaller than a^{-1} due to the presence of the potential $v(\mathbf{k})$.

So we can choose the diagrams for self-energy parts with the minimal number of these factors. They are

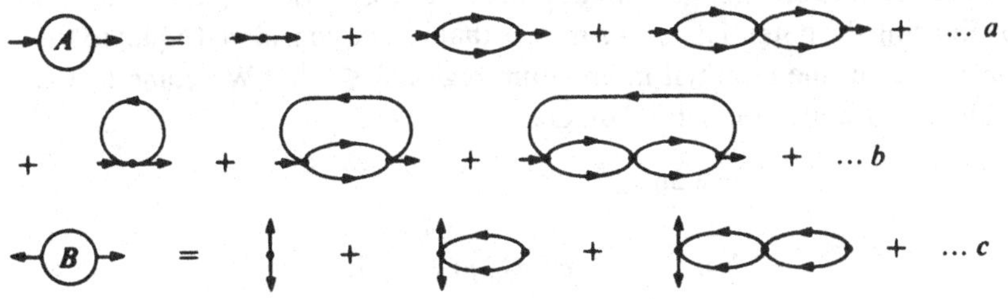

(6.3)

The sets a, c are proportional to $\alpha^2 = \rho_0$, every diagram of the b-set contains one cycle of normal lines. In all sets a, b, c we are dealing with the following set of four-tail parts:

(6.4)

The summation of this set is equivalent to the solution of the equation

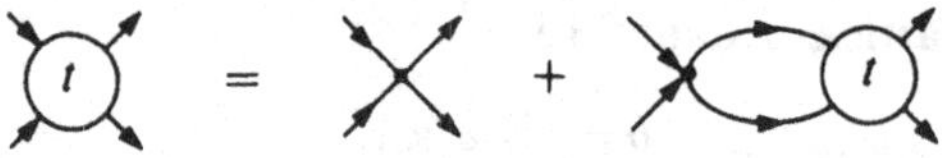

(6.5)

The main contribution into integrals over momenta comes from $k \sim a^{-1} \gg T^{1/2} \gg \lambda^{1/2}$. Thus we can neglect λ in the Green's function

$$\left(\mathrm{i}\omega - \frac{k^2}{2m} + \lambda \right)^{-1} \to (\mathrm{i}\omega - k^2)^{-1} \quad (2m = 1)$$

and convert the sum over ω into an integral. As a result, we find

$$\approx t(\tfrac{1}{2}(\mathbf{k}_1 - \mathbf{k}_2), \tfrac{1}{2}(\mathbf{k}_3 - \mathbf{k}_4), \mathrm{i}\omega_1 + \mathrm{i}\omega_2 - \tfrac{1}{2}(\mathbf{k}_1 + \mathbf{k}_2)^2)$$
$$+ t(\tfrac{1}{2}(\mathbf{k}_1 - \mathbf{k}_2), \tfrac{1}{2}(\mathbf{k}_4 - \mathbf{k}_3), \mathrm{i}\omega_1 + \mathrm{i}\omega_2 - \tfrac{1}{2}(\mathbf{k}_1 + \mathbf{k}_2)^2), \qquad (6.6)$$

where $t(\mathbf{k}_1, \mathbf{k}_2, z)$ is the so-called t-matrix, which describes the scattering of two particles in vacuum. The equation for the t-matrix is

$$t(\mathbf{k}_1, \mathbf{k}_2, z) + (2\pi)^{-3} \int \frac{\mathrm{d}^3 k_3 v(\mathbf{k}_1 - \mathbf{k}_3) t(\mathbf{k}_3, \mathbf{k}_2, z)}{k_3^2 - z} = v(\mathbf{k}_1 - \mathbf{k}_2). \qquad (6.7)$$

The function (6.6) will be of interest mainly in the region $k \ll a^{-1}, \omega \ll a^{-2}$. For such $(\mathbf{k}, \omega)$ we may consider (6.6) equal to its value for $\omega = 0$, $\mathbf{k} = 0$:

$$\approx 2t(0, 0, 0) = 2t_0. \qquad (6.8)$$

Now it is easy to see that the summation of the a, b, c series in (6.3) is reduced to replacing the potential v by a t-matrix that can be considered equal to the constant t_0 in the essential momentum region $k \lesssim T^{1/2}$. We come to the following formulae for sets of diagrams a, b, c (6.3):

$$a \approx 2t_0\rho_0, \quad c \approx t_0\rho_0,$$

$$b \approx -\frac{2t_0}{\beta V} \sum_p e^{\mathrm{i}\omega\varepsilon} G_0(p) = 2t_0\rho_1. \qquad (6.9)$$

Here ρ_1 is the density of noncondensate particles, which can be calculated (in the first approximation) by using the Green's function $G_0(p) = (i\omega - k^2)^{-1}$. We then deduce the following formula for ρ_1:

$$\rho_1 = \frac{1}{V}\sum_{\mathbf{k}\neq 0}(e^{\beta k^2}-1)^{-1} \approx (2\pi)^{-3}\int d^3k(e^{\beta k^2}-1)^{-1}$$

$$= T^{3/2}(2\pi)^{-3}4\pi\int_0^\infty (e^{x^2}-1)^{-1}x^2 dx = \zeta(3/2)(4\pi)^{-3/2}T^{3/2}. \quad (6.10)$$

Here $\zeta(3/2)$ is the Riemann ζ-function $\zeta(s)$ for $s = 3/2$. For the self-energy parts A, B we get

$$A(p) = 2t_0\rho_0 + 2t_0\rho_1 = 2t_0\rho,$$

$$B(p) = t_0\rho_0. \quad (6.11)$$

Substituting this into the Hugengoltz–Pines condition (5.21) we find a nontrivial solution for $\rho_0 = |\alpha|^2$:

$$2t_0\rho_0 + 2t_0\rho_1 - t_0\rho_0 - \lambda = 0 \quad (6.12)$$

or

$$\rho_0 = \frac{\lambda}{t_0} - 2\zeta(3/2)(4\pi)^{-3/2}T^{3/2} = \frac{\Lambda}{t_0}, \quad (6.13)$$

where

$$\Lambda = \lambda - 2t_0\zeta(3/2)(4\pi)^{-3/2}T^{3/2}. \quad (6.14)$$

We will suppose t_0 positive. This is so for the repulsive potential. We see that the positive solution ρ_0 exists if and only if $\Lambda > 0$. That is why the equation

$$\Lambda = 0 \quad (6.15)$$

gives us the curve of the phase transition in the λ–T plane. For $\Lambda > 0$ the system is in superfluid state, and for $\Lambda < 0$ we have the normal state.

For the normal state we have

$$\rho_0 = 0, B = 0, A = 2t_0\rho_1 = 2t_0\zeta(3/2)(4\pi)^{-3/2}T^{3/2}$$

$$G(p) = (i\omega - k^2 + \lambda - A)^{-1} = (i\omega - k^2 + \Lambda)^{-1}. \quad (6.16)$$

For the superfluid state we obtain

$$A = 2t_0(\rho_0 + \rho_1),\ A - \lambda = \Lambda,\ B = \Lambda, \quad (6.17)$$

and we arrive at the following formulae for the normal and anomalous

Green's functions:

$$G(p) = \frac{i\omega + k^2 + \Lambda}{\omega^2 + k^4 + 2\Lambda k^2}$$

$$G_1(p) = \frac{\Lambda}{\omega^2 + k^4 + 2\Lambda k^2}. \tag{6.18}$$

We are now in a position to determine the so-called energy spectrum of quasiparticles, by examining the poles of Green's functions after the analytical continuation $i\omega \to E$. From (6.16) and (6.18) we find

$$E = (k^4 + 2\Lambda k^2)^{1/2}, \quad \Lambda > 0$$

$$E = k^2 - \Lambda = k^2 + |\Lambda|, \quad \Lambda < 0. \tag{6.19}$$

The first of these formulae is the well-known Bogoliubov spectrum. It behaves like $k^2 + \Lambda$ for $k^2 \gg \Lambda$ and like ck, $c = (2\Lambda)^{1/2}$, for $k^2 \ll \Lambda$. At small momenta we have a linear phonon-like spectrum. The formulae obtained for the normal Green's functions of normal and superfluid states, (6.16) and (6.18), turn one into the other on the phase transition curve $\Lambda = 0$. We must remember, however, that these formulae break down for $\Lambda = 0$. They are very good approximations only outside a narrow strip along the $\Lambda = 0$ curve in the λ–T plane.

The considerations determining the main diagram series for the self-energy parts of Green's functions A, B are also applicable if we want to find the main diagram series for the pressure p. The starting formula is

$$p = p_0 + C_0 + D, \tag{6.20}$$

where p_0 is the pressure of the ideal Bose gas, C_0 is the constant ($C_0 = \lambda|\alpha|^2 - \frac{1}{2}v(0)|\alpha|^4$) determined by (5.5), and D is the sum of all connected (vacuum) diagrams. The main contribution into p comes from the diagrams with the minimal number of the above-mentioned factors (1) and (2):

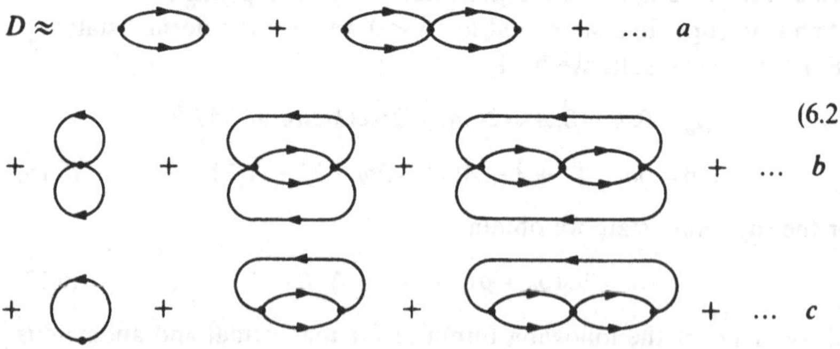

(6.21)

These diagrams contain either two vertices of the second-order or two independent loops, or one independent loop with either one second-order vertex or two third-order vertices. These diagrams form three series. In all the series we deal with the same set of four-tail parts (6.4), as for A, B above. The summation of this set is equivalent to replacing the potential v by the t-matrix. The result is that the sum of the term $-\frac{1}{2}\rho_0^2 v(0)$ in C_0 and of series (a) is equal to $-\frac{1}{2}t_0\rho_0^2$, series (b) is equal to $-t_0\rho_1^2$, and for series (c) we obtain the expression $-2t_0\rho_0\rho_1$, where ρ_1 is the density of noncondensate particles as in (6.10).

The pressure of the ideal gas is

$$-\frac{1}{\beta V}\sum_{\mathbf{k}} \ln(1-\exp-\beta(k^2-\lambda))$$

$$\approx -\frac{1}{\beta V}\sum_{\mathbf{k}} \ln(1-e^{-\beta k^2})+\frac{\lambda}{V}\sum_{\mathbf{k}}(e^{\beta k^2}-1)^{-1}$$

$$=\zeta(5/2)(4\pi)^{-3/2}T^{5/2}+\lambda\zeta(3/2)(4\pi)^{-3/2}T^{3/2}. \qquad (6.22)$$

Inserting the above expressions into (6.20), we get

$$p=\zeta(5/2)(4\pi)^{-3/2}T^{5/2}+\lambda\zeta(3/2)(4\pi)^{-3/2}T^{3/2}$$

$$+\lambda\rho_0-\frac{t_0}{2}\rho_0^2-t_0\rho_1^2-2t_0\rho_0\rho_1. \qquad (6.23)$$

Substituting ρ_0 and ρ_1 from (6.10) and (6.13) into (6.23) we obtain the expression for the pressure p in terms λ and T

$$p=\zeta(5/2)(4\pi)^{-3/2}T^{5/2}+\frac{\lambda^2}{2t_0}-\lambda\zeta(3/2)(4\pi)^{-3/2}T^{3/2}$$

$$+t_0\zeta^2(3/2)(4\pi)^{-3}T^3. \qquad (6.24)$$

Here the first term is the leading one, the others are of the same order, and their relation to the leading term is of the order

$$t_0T^{1/2}\sim t_0\rho^{1/3}\ll 1.$$

Differentiating (6.24) with the respect to λ, we obtain an expression for the full density ρ:

$$\rho=\frac{\partial p}{\partial\lambda}=\frac{\lambda}{t_0}-\zeta(3/2)(4\pi)^{-3/2}T^{3/2}. \qquad (6.25)$$

The same result can be derived by taking the sum $\rho_0+\rho_1$ according to (6.10) and (6.13).

Putting $\rho_0 = 0$, $\rho_1 = \zeta(3/2)(4\pi)^{-3/2}T^{3/2}$ into (6.23), we obtain the expression for the pressure in the normal state

$$p = \zeta(5/2)(4\pi)^{-3/2}T^{5/2} + \lambda\zeta(3/2)(4\pi)^{-3/2}T^{3/2} - t_0\zeta^2(3/2)(4\pi)^{-3}T^3. \quad (6.26)$$

Differentiating (6.26) with respect to λ, we get the density in the normal state

$$\rho = \rho_1 = \frac{\partial p}{\partial \lambda} = \zeta(3/2)(4\pi)^{-3/2}T^{3/2}. \quad (6.27)$$

We can now consider the extra diagrams. For instance, we can calculate the second-order approximation to the self-energy parts. It turns out that if we take into account the second-order diagrams the poles of the Green's functions move from the real axis of the E-plane to the lower half-plane. It means that quasiparticles would have a finite lifetime.

Now, it is important that the second-order diagrams become singular at small k, ω. For instance, let us consider the diagram

δA =  (6.28)

where 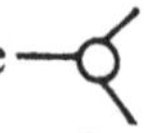 and 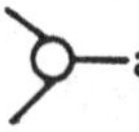 are vertices which are equal to $2\alpha t_0$ (they can be obtained from the bare ones  and  by replacing v by t_0). The expression corresponding to (6.28) is

$$-\frac{1}{r\beta V}\sum_{\mathbf{k}_1,\omega_1}(2\alpha t_0)^2 G(p_1)G(p_2), \quad (6.29)$$

where $p_1 + p_2 = p$, r is the order of the symmetry group. Let us put $\omega = 0$ and consider only one term, $\omega_1 = 0$, in the sum over ω_1. We obtain

$$-\frac{4\rho_0 t_0^2}{r\beta V}\sum_{k_1} G(0, k_1)G(0, k - k_1). \quad (6.30)$$

According to (6.19) for small $k(k^2 \ll \Lambda)$ we have

$$G(0, k) \approx -\frac{1}{2k^2}, \quad (6.31)$$

and the following result for (6.30):

$$-\frac{\rho_0 t_0^2 T}{r(2\pi)^3}\int\frac{d^3k_1}{k_1^2(\mathbf{k} - \mathbf{k}_1)^2} = -\frac{\rho_0 t_0^2 T}{8rk} = -\frac{\Lambda t_0 T}{8rk}. \quad (6.32)$$

Let us compare it with the self-energy part in the first approximation which is of the same order as Λ. So we have

$$\frac{\delta A}{A} \sim \frac{t_0 T}{k}. \tag{6.33}$$

We see that δA becomes infinite for $k \to 0$, and is of the same order as A for $k \sim t_0 T$.

It is clear that when physical quantities are evaluated, divergences must cancel out. Nevertheless the appearance of singularities in separate diagrams of the perturbation theory is unwanted.

A reformulation of the perturbation theory that eliminates the singularities is highly desirable. Such a reformulation in the functional integral formalism is investigated in the following section.

Knowing Green's functions (6.16) and (6.18) we can calculate the noncondensate density ρ_1 with more accuracy. If $\Lambda < 0$, ρ_1 is equal to the full density ρ, and we have ($\varepsilon \to +0$):

$$\begin{aligned}\rho = \rho_1 &= -\frac{1}{\beta V}\sum_p e^{i\omega\varepsilon}(i\omega - k^2 - |\Lambda|)^{-1} \\ &= \frac{1}{V}\sum_{\mathbf{k}}(\exp\beta(k^2 + |\Lambda|) - 1)^{-1} \\ &\approx (4\pi)^{-3/2}\zeta(3/2)T^{3/2} - (4\pi)^{-1}T|\Lambda|^{1/2}.\end{aligned} \tag{6.34}$$

If $\Lambda > 0$, we get

$$\begin{aligned}\rho_1 &= \frac{1}{\beta V}\sum_p e^{i\omega\varepsilon}(i\omega + k^2 + \Lambda)(\omega^2 + k^4 + 2\Lambda k^2)^{-1} \\ &= (2V)^{-1}\sum_{\mathbf{k}}\left(\frac{k^2+\Lambda}{2\varepsilon(\mathbf{k})}\operatorname{cotanh}\tfrac{1}{2}\beta\varepsilon(\mathbf{k}) - 1\right) \\ &\approx (4\pi)^{-3/2}\zeta(3/2)T^{3/2} - (8\pi)^{-1}T(2\Lambda)^{1/2},\end{aligned} \tag{6.35}$$

where $\varepsilon(\mathbf{k})$ is the Bogoliubov energy spectrum.

It is more difficult to calculate the condensate density with more accuracy. One can do it considering (5.21) and taking into account contributions of the second-order diagrams (like (6.28)) into $A(p)$, $B(p)$. The result is

$$\rho_0 = \frac{\Lambda}{t_0} + \frac{3}{8\pi}T(2\Lambda)^{1/2}. \tag{6.36}$$

Adding (6.35) and (6.36) we obtain the full density

$$\rho = \rho_0 + \rho_1 = \frac{\lambda}{t_0} - (4\pi)^{-3/2}\zeta(3/2)T^{3/2} + (4\pi)^{-1}T(2\Lambda)^{1/2}. \tag{6.37}$$

Then using equation $\rho = \partial p/\partial \lambda$, we can find the correction to the formula (6.24) for the pressure p, corresponding to the correction $(4\pi)^{-1}T(2\Lambda)^{1/2}$ to ρ in (6.37). This correction is

$$\Delta p = (12\pi)^{-1}T(2\Lambda)^{3/2}. \tag{6.38}$$

The corresponding correction to p in the normal state (6.25) is equal to

$$\Delta p = (6\pi)^{-1}T|\Lambda|^{3/2}. \tag{6.39}$$

In order to conclude this section we shall consider the phenomenon of superfluidity. Let us take the Bose system in the coordinate system moving with the velocity $\mathbf{v}$. In such a case the system is described by the functional

$$S + \int_0^\beta d\tau \int dx \frac{i}{2m}(\nabla\bar{\psi}\psi - \bar{\psi}\nabla\psi), \tag{6.40}$$

where S is the functional as in (2.3).

In the normal state the mean values of the number of particles and momentum are related by

$$\mathbf{K} = m\mathbf{v}N = m\mathbf{v}V\rho, \tag{6.41}$$

which means that the whole system moves with the velocity $\mathbf{v}$. When the condensate appears, the system may be regarded as consisting of two components: the normal one ρ_n moving with velocity $\mathbf{v}$ and the superfluid one $\rho_s = \rho - \rho_n$ moving with the velocity of condensate $\mathbf{v}_0$, which may be different from $\mathbf{v}$. For example, for the condensate at rest we obtain a smaller value

$$\mathbf{K} = m\mathbf{v}V\rho_n \tag{6.42}$$

instead of (6.41).

In the first approximation the Green's function in the moving coordinate system can be obtained from (6.18) by the shift $i\omega \to i\omega + (\mathbf{v}, \mathbf{k})$, and we have

$$G(p) = \frac{i\omega + k^2 + (\mathbf{v}, \mathbf{k}) + \Lambda}{(i\omega + k^2 + (\mathbf{v}, \mathbf{k}) + \Lambda)(i\omega + (\mathbf{v}, \mathbf{k}) - k^2 - \Lambda) + \Lambda^2}. \tag{6.43}$$

The energy spectrum

$$E(\mathbf{k}) = -(\mathbf{v}, \mathbf{k}) + (k^4 + 2\Lambda k^2)^{1/2} \tag{6.44}$$

is stable ($E(\mathbf{k}) > 0$) if $v < (2\Lambda)^{1/2}$. Knowing Green's function (6.37), we can find the momentum mean value

$$\begin{aligned} \mathbf{K} &= -\beta^{-1}\sum_p e^{i\omega\varepsilon}\mathbf{k}G(p) \\ &\approx m\mathbf{v}V[(4\pi)^{-3/2}\zeta(3/2)T^{3/2} - (6\pi)^{-1}T(2\Lambda)^{1/2}]. \end{aligned} \tag{6.45}$$

Comparing this with (6.42), we get

$$\rho_n = (4\pi)^{-3/2}\zeta(3/2)T^{3/2} - (6\pi)^{-1}T(2\Lambda)^{1/2},$$

$$\rho_s = \rho - \rho_n = \frac{\Lambda}{t_0} + \frac{5}{12\pi}T(2\Lambda)^{1/2}. \qquad (6.46)$$

If we compare (6.46) with (6.35) and (6.36) we find that the condensate density ρ_0 in the first approximation is equal to the superfluid density $\rho_s(\rho_0 \approx \rho_s \approx \Lambda/t_0)$, and the noncondensate density ρ_1 is approximately equal to $\rho_n(\rho_1 \approx \rho_n \approx \lambda/t_0 - (4\pi)^{-3/2}\zeta(3/2)T^{3/2})$. But if we take into account corrections $\sim T\Lambda^{1/2}$ we can see that ρ_0 differs from ρ_s and ρ_1 differs from ρ_n.

7
The modified perturbative scheme for superfluid Bose systems

In the two previous sections we developed a perturbation theory which takes into account Bose condensation in a system of Bose particles. This theory is useful for many purposes, but unfortunately it has infrared singularities. Here we will present a modified perturbative scheme which was developed in the functional integral framework in order to avoid infrared singularities at small momenta $p = (\mathbf{k}, \omega)$. As was explained in section 4, the main idea is to perform functional integration in two steps. First, one integrates over 'fast' fields. When integrating over 'slow' variables, it turns out to be convenient to introduce new variables of density-phase type. The perturbative scheme developed in terms of Green's functions in these new variables is then completely free of infrared singularities. We define the 'slow' part $\psi_0(\mathbf{x}, \tau)$ and the 'fast' part $\psi_1(\mathbf{x}, \tau)$ of the ψ field in the following way:

$$\psi_0(\mathbf{x}, \tau) = (\beta V)^{-1/2} \sum_{\omega, |\mathbf{k}| < k_0} e^{i(\omega\tau + \mathbf{k}\mathbf{x})} a(p),$$
$$\psi_1(\mathbf{x}, \tau) = (\beta V)^{-1/2} \sum_{\omega, |\mathbf{k}| > k_0} e^{i(\omega\tau + \mathbf{k}\mathbf{x})} a(p). \tag{7.1}$$

Here the parameter k_0 separates the 'small' k from the 'large' ones. We can estimate only the order of k_0. This order of magnitude depends on the specific Bose system under consideration.

Performing integration over fast variables we obtain the following result:

$$\int e^S D\bar{\psi}_1 D\psi_1 = \exp S_{\text{eff}}[\psi_0, \bar{\psi}_0]. \tag{7.2}$$

Here S_{eff} or S_{h} (hydrodynamic) has the meaning of the effective action functional, which describes 'slow' fields.

In order to integrate over the $\psi_0, \bar{\psi}_0$ fields we introduce two new variables: the density $\rho(x, \tau)$ and the phase $\varphi(x, \tau)$. These variables are defined by

$$\psi_0(\mathbf{x}, \tau) = (\rho(\mathbf{x}, \tau))^{1/2} e^{i\varphi(\mathbf{x}, \tau)}$$
$$\bar{\psi}_0(\mathbf{x}, \tau) = (\rho(\mathbf{x}, \tau))^{1/2} e^{-i\varphi(\mathbf{x}, \tau)}. \tag{7.3}$$

We have

$$d\bar{\psi}_0(\mathbf{x},\tau)\,d\psi_0(\mathbf{x},\tau) = d\rho(\mathbf{x},\tau)\,d\varphi(\mathbf{x},\tau). \tag{7.4}$$

Let us insert $\psi_0 = \bar{\psi}_0 = \rho_0^{1/2} = \text{const}$ into $S_{\text{eff}}[\psi_0, \bar{\psi}_0]$. Then S_{eff} becomes a function of the single variable ρ_0 and we may consider the equation

$$\frac{\partial S_{\text{eff}}}{\partial \rho_0} = 0. \tag{7.5}$$

If at the first step we integrate over all fields except constant ones we would have $S_{\text{eff}} = -\beta\Omega = \beta V p$. Then (7.5) is equivalent to $\partial p/\partial\rho_0 = 0$ (5.26) which defines ρ_0 below the Bose-condensation point. But since we do not integrate over all fields, S_{eff} depends on the parameter k_0. The solution ρ_0 in (7.5) (if it exists) then also depends on k_0. We will call $\rho_0(k_0)$ the bare ('initial') condensate density.

The value $\rho_0(k_0)$ is close to the real condensate density ρ_0 for three-dimensional systems ($d = 3$). In this case $\rho_0(k_0) \to \rho_0$, as $k_0 \to 0$. On the other hand in the cases when $d = 2$ (small $T \neq 0$) or $d = 1$ and $T = 0$, $\rho_0(k_0) \to 0$ as $k_0 \to 0$.

It turns out, however, that the existence of the initial condensate $\rho_0(k_0)$ is sufficient for the system to be superfluid. We can demonstrate the possibility that the system may be superfluid without Bose condensation in the cases $d = 2$, $T \neq 0$ or $d = 1$, $T = 0$. We will discuss this phenomenon in detail below after a discussion of the explicit form of $S_{\text{eff}}[\psi_0, \bar{\psi}_0]$.

Assuming that the solution $\rho_0(k_0)$ of (7.5) exists we introduce a new variable

$$\pi(\mathbf{x},\tau) = \rho(\mathbf{x},\tau) - \rho_0(k_0), \tag{7.6}$$

which plays the role of the fluctuation of $\rho(x,\tau)$ near its average value $\rho_0(k_0)$. Now let us consider the starting action functional S (2.3) and write it explicitly in terms of the fast and slow fields ψ_0, ψ_1. We obtain

$$\begin{aligned} S = {} & \int_0^\beta d\tau \int d\mathbf{x}(\bar{\psi}_0\partial_\tau\psi_0 - (2m)^{-1}\nabla\bar{\psi}_0\nabla\psi_0 + \lambda\bar{\psi}_0\psi_0) \\ & + \int_0^\beta d\tau \int d\mathbf{x}(\bar{\psi}_1\partial_\tau\psi_1 - (2m)^{-1}\nabla\bar{\psi}_1\nabla\psi_1 + \lambda\bar{\psi}_1\psi_1) \\ & - \frac{1}{2}\int_0^\beta d\tau \int d\mathbf{x}\, d\mathbf{y}\, u(\mathbf{x}-\mathbf{y})[\bar{\psi}_0(1)\bar{\psi}_0(2)\psi_0(2)\psi_0(1) \\ & + 2\bar{\psi}_1(1)\bar{\psi}_0(2)\psi_0(2)\psi_0(1) + 2\bar{\psi}_0(1)\bar{\psi}_0(2)\psi_0(2)\psi_1(1) \\ & + 2\bar{\psi}_0(1)\psi_0(1)\bar{\psi}_1(2)\psi_1(2) + 2\bar{\psi}_0(1)\bar{\psi}_1(2)\psi_0(2)\psi_1(1) \end{aligned}$$

$$+\bar{\psi}_0(1)\bar{\psi}_0(2)\psi_1(2)\psi_1(1)+\bar{\psi}_1(1)\bar{\psi}_1(2)\psi_0(2)\psi_0(1)$$
$$+2\bar{\psi}_0(1)\bar{\psi}_1(2)\psi_1(2)\psi_1(1)+2\bar{\psi}_1(1)\bar{\psi}_1(2)\psi_1(1)\psi_0(1)$$
$$+\bar{\psi}_1(1)\bar{\psi}_1(2)\psi_1(2)\psi_1(1)] \tag{7.7}$$

with the shorthand notation $\psi_0(1)=\psi_0(\mathbf{x},\tau)$, $\psi_0(2)=\psi_0(\mathbf{y},\tau)$, $\psi_1(1)=\psi_1(\mathbf{x},\tau)$, $\psi_1(2)=\psi_1(\mathbf{y},\tau)$. In (7.7) we have taken into account that the crossing terms $\bar{\psi}_1\psi_0$ or $\bar{\psi}_0\psi_1$ vanish when integrated over the x-variable.

Now it is suitable to perform the following phase transformation:

$$\psi_1(\mathbf{x},\tau)\to\psi_1(\mathbf{x},\tau)e^{i\varphi(\mathbf{x},\tau)},\quad \bar{\psi}_1(\mathbf{x},\tau)\to\bar{\psi}_1(\mathbf{x},\tau)e^{i\varphi(\mathbf{x},\tau)}, \tag{7.8}$$

where $\varphi(\mathbf{x},\tau)$ is the phase of the 'slow' field $\psi_0(\mathbf{x},\tau)$. After this transformation the fourth-order form of $\psi_0,\psi_1,\bar{\psi}_0,\bar{\psi}_1$ in (7.7) does not depend on the φ-variable at all, and we may replace both $\psi_0(\mathbf{x},\tau)$, and $\bar{\psi}_0(\mathbf{x},\tau)$ by $(\rho(\mathbf{x},\tau))^{1/2}$ in this form. The quadratic form of the $\psi_1,\bar{\psi}_1$ variables in (7.7) becomes

$$\int_0^\beta d\tau\int dx(\bar{\psi}_1\partial_\tau\psi_1-(2m)^{-1}\nabla\bar{\psi}_1\nabla\psi_1+(2m)^{-1}i(\bar{\psi}_1\nabla\psi_1-\nabla\bar{\psi}_1\psi_1)\nabla\varphi$$
$$+\bar{\psi}_1\psi_1(\lambda-(2m)^{-1}(\nabla\varphi)^2+i\partial_\tau\varphi)) \tag{7.9}$$

after the transformation (7.8). We see that after the phase transformation (7.8) the action functional S depends on the derivatives $\partial_\tau\varphi,\nabla\varphi$ of the phase variable φ, but not on the phase φ itself.

Now we can build up the perturbation theory for integration over fast fields. We regard the quadratic form

$$\int_0^\beta d\tau\int dx(\bar{\psi}_1\partial_\tau\psi_1-(2m)^{-1}\nabla\bar{\psi}_1\nabla\psi_1+\lambda\bar{\psi}_1\psi_1) \tag{7.10}$$

as the free field theory action. All other forms are considered as perturbations. We arrive at a perturbation theory which differs from that considered in section 5 in the following points:

(1) The 'anomalous' vertices (depending on $\alpha,\bar{\alpha}$ in the simplest scheme) now depend on $(\rho(\mathbf{x},\tau))^{1/2}$.

(2) Some new vertices of the second order emerge. They are due to the terms in (7.9) containing $\partial_\tau\varphi$ or $\nabla\varphi$.

(3) All sums (integrals) over momenta are cut at the lower limit k_0.

Now the functional S_{eff} may be defined as the sum of all vacuum diagrams of this new perturbation theory. We can say that the system of fast particles interacts with the following slowly varying fields: the field on nonhomogeneous condensate $\rho(\mathbf{x},\tau)$, the field of nonhomogeneous chemical potential

$$\lambda(\mathbf{x},\tau)=\lambda+i\partial_\tau\varphi(\mathbf{x},\tau)-(2m)^{-1}(\nabla\varphi(\mathbf{x},\tau))^2 \tag{7.11}$$

and the velocity field

$$\mathbf{v}(x,\tau) = m^{-1}\nabla\varphi(\mathbf{x},\tau). \tag{7.12}$$

Let us consider the expansion of $S_{\rm eff}$ into the functional series of variables $\pi(\mathbf{x},\tau) = \rho(\mathbf{x},\tau) - \rho_0(k_0)$, $\delta\lambda(\mathbf{x},\tau) = \lambda(\mathbf{x},\tau) - \lambda$, $\mathbf{v}(\mathbf{x},\tau)$. It is clear that only the first few terms of this series are essential for small values of k_0. We can obtain the first approximation for this expansion by using

$$S_{\rm eff} \approx \int_0^\beta d\tau \int dx\, p(\lambda(\mathbf{x},\tau), \rho(\mathbf{x},\tau), \mathbf{v}(\mathbf{x},\tau)), \tag{7.13}$$

where $p(\lambda,\rho,\mathbf{v})$ is the pressure in the homogeneous system with chemical potential λ, condensate density ρ and velocity field v.

Let us consider the case of very small temperatures (or large β). It can be shown that in that case $p(\lambda,\rho,\mathbf{v})$ does not depend on v, and we can use the Taylor series expansion of $p(\lambda,\rho)$

$$\begin{aligned} p(\lambda+\delta\lambda, \rho_0(k_0)+\pi, 0) &= p(\lambda,\rho_0(k_0),0) + p_\lambda\delta\lambda + p_{\rho_0}\pi \\ &\quad + \tfrac{1}{2}p_{\lambda\lambda}(\delta\lambda)^2 + p_{\lambda\rho_0}(\delta\lambda)\pi + \tfrac{1}{2}p_{\rho_0\rho_0}\pi^2 + \cdots \end{aligned} \tag{7.14}$$

Here $p_{\rho_0} = 0$ due to the choice of $\rho_0(k_0)$. Thus we come to the following expression for $S_{\rm eff}$:

$$\begin{aligned} S_{\rm eff} &= \beta V p(\lambda, \rho_0(k_0), 0) \\ &\quad + \int d\tau\, dx\, \{p_\lambda(i\partial_\tau\varphi - (2m)^{-1}(\nabla\varphi)^2) + \tfrac{1}{2}p_{\lambda\lambda}(i\partial_\tau\varphi - (2m)^{-1}(\nabla\varphi)^2)^2 \\ &\quad + p_{\lambda\rho_0}(i\partial_\tau\varphi - (2m)^{-1}(\nabla\varphi)^2)\pi + \tfrac{1}{2}p_{\rho_0\rho_0}\pi^2 + \cdots\}. \end{aligned} \tag{7.15}$$

The quadratic form in this expansion is

$$\int d\tau\, dx\left(-\frac{p_\lambda}{2m}(\nabla\varphi)^2 - \tfrac{1}{2}p_{\lambda\lambda}(\partial_\tau\varphi)^2 + ip_{\lambda\rho_0}\pi\partial_\tau\varphi + \tfrac{1}{2}p_{\rho_0\rho_0}\pi^2\right). \tag{7.16}$$

The coefficients in (7.16) are expressed in terms of thermodynamical functions of the system. Equation (7.16) gives us the exact expression of the quadratic part of $S_{\rm eff}$ for any superfluid Bose system at low temperature.

We can also take into account corrections to (7.16). For a Bose gas model the main corrections come from the kinetic energy form of the ψ_0 field

$$\begin{aligned} &-\int d\tau\, dx (2m)^{-1}\nabla\bar{\psi}_0\nabla\psi_0 \\ &= -\int \frac{d\tau\, dx}{2m}\left[(\rho_0+\pi)(\nabla\varphi)^2 + \frac{(\nabla\pi)^2}{4(\rho_0+\pi)}\right]. \end{aligned} \tag{7.17}$$

The first term $\rho_0(\nabla\varphi)^2$ contributes to the quadratic form (7.16). So we only have to take into account $\pi(\nabla\varphi)^2$ and $(\nabla\pi)^2/4(\rho_0+\pi)$, where we may further neglect π in the denominator.

Hence, we will consider the following action functional:

$$\int \mathrm{d}\tau\,\mathrm{d}x\left(\mathrm{i}p_{\lambda\rho_0}\pi\partial_\tau\varphi - \frac{p_\lambda}{2m}(\nabla\varphi)^2 - \tfrac{1}{2}p_{\lambda\lambda}(\partial_\tau\varphi)^2 + \tfrac{1}{2}p_{\rho_0\rho_0}\pi^2 - \frac{(\nabla\pi)^2}{8m\rho_0} - \frac{\pi(\nabla\varphi)^2}{2m}\right) \tag{7.18}$$

for which a new perturbative scheme can be developed by taking the quadratic form (7.16) as the free field theory action. Let us rewrite (7.18) in terms of the Fourier coefficients $\varphi(p)$, $\pi(p)$ of the φ, π fields

$$\tfrac{1}{2}\sum_p\left[-\left(\frac{p_\lambda}{m}k^2 + p_{\lambda\lambda}\omega^2\right)\varphi(p)\varphi(-p) - 2p_{\lambda\rho_0}\omega\varphi(p)\pi(-p) + \left(p_{\rho_0\rho_0} - \frac{k^2}{2m\rho_0}\right)\pi(p)\pi(-p)\right] + (\beta V)^{-1/2}\sum_{p_1+p_2+p_3=0}\frac{(\mathbf{k}_1\mathbf{k}_2)}{2m}\varphi(p_1)\varphi(p_2)\pi(p_3). \tag{7.19}$$

The elements of the diagram techniques are

$\langle\varphi(p)\varphi(-p)\rangle_0$

$\langle\pi(p)\pi(-p)\rangle_0$

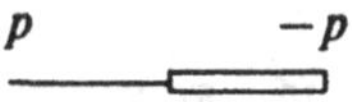

$\langle\varphi(p)\pi(-p)\rangle_0$

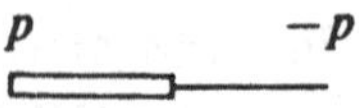

$\langle\pi(p)\varphi(-p)\rangle_0$

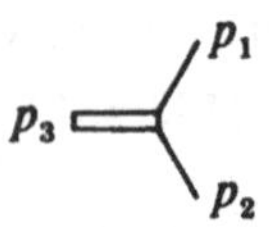

$$\frac{(\mathbf{k}_1\mathbf{k}_2)}{m}. \tag{7.20}$$

Here we denote the φ field by a single line, and the π field by a double line.

The Green functions $\langle\varphi(p)\varphi(-p)\rangle_0$, $\langle\varphi(p)\pi(-p)\rangle_0$, $\langle\pi(p)\varphi(-p)\rangle_0$,

$\langle \pi(p)\pi(-p)\rangle_0$ may be considered as the matrix elements of the matrix $G(p)$ whose inverse is

$$G^{-1}(p) = \begin{pmatrix} \frac{p_\lambda}{m}k^2 + p_{\lambda\lambda}\omega^2, & -p_{\lambda\rho_0}\omega \\ p_{\lambda\rho_0}\omega, & -p_{\rho_0\rho_0} + \frac{k^2}{4m\rho_0} \end{pmatrix}. \tag{7.21}$$

For a Bose gas model the pressure at small T is given by

$$p \approx \lambda\rho_0 - \frac{t_0}{2}\rho_0^2. \tag{7.22}$$

This formula may be obtained from (6.23) by putting $T = 0$, $\rho_1 = 0$. From (7.22) we find:

$$p_\lambda = \rho \approx \rho_0, \quad p_{\lambda\lambda} = 0, \quad p_{\lambda\rho_0} = 1, \quad p_{\rho_0\rho_0} = -t_0, \tag{7.23}$$

and we come to the following Green's functions:

$$\langle \varphi(p)\varphi(-p)\rangle_0 = \frac{t_0 + \dfrac{k^2}{4m\rho_0}}{\omega^2 + \varepsilon^2(\mathbf{k})},$$

$$\langle \pi(p)\pi(-p)\rangle_0 = \frac{\dfrac{\rho_0}{m}k^2}{\omega^2 + \varepsilon^2(\mathbf{k})},$$

$$\langle \varphi(p)\pi(-p)\rangle_0 = -\langle \pi(p)\varphi(-p)\rangle_0 = \frac{\omega}{\omega^2 + \varepsilon^2(\mathbf{k})}, \tag{7.24}$$

where

$$\varepsilon^2(\mathbf{k}) = \left(\frac{k^2}{2m}\right)^2 + \frac{t_0\rho_0}{m}k^2 \tag{7.25}$$

is the square of Bogoliubov spectrum.

It is now easy to show that the perturbation theory outlined above has no infrared divergences. The singularities of the $\langle\varphi\varphi\rangle_0$ or $\langle\varphi\pi\rangle_0$ ($\langle\pi\varphi\rangle_0$) Green's functions are cancelled out by the $(\mathbf{k}_1\mathbf{k}_2)/m$ factors corresponding to the vertices attached to the $\langle\varphi\varphi\rangle_0$ or $\langle\varphi\pi\rangle_0$($\langle\pi\varphi\rangle_0$) line:

(7.26)

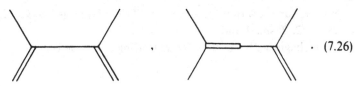

Obviously, that cancellation of singularities takes place for two-dimensional or one-dimensional systems as well. The only thing we need is the existence of $\rho_0(k_0)$ and the possibility of the functional expansion of S_{eff}.

Let us show that the mean $\langle\psi(\mathbf{x},\tau)\bar{\psi}(\mathbf{x}',\tau')\rangle$ behaves like $r^{-\alpha}$ for two-dimensional systems at low temperature and for one-dimensional ones at $T=0$, if $\rho_0(k_0)$ exists and if S_{eff} has the same form as in (7.19).

First of all we have

$$\langle\psi(\mathbf{x},\tau)\bar{\psi}(\mathbf{x}',\tau')\rangle = \langle\psi_0(\mathbf{x},\tau)\bar{\psi}_0(\mathbf{x}',\tau')\rangle + \langle\psi_1(\mathbf{x},\tau)\bar{\psi}_1(\mathbf{x}',\tau')\rangle. \quad (7.27)$$

The second term on the right-hand side of (7.27) is much smaller than the first one for large $r=|\mathbf{x}-\mathbf{x}'|$. So we will consider the average

$$\begin{aligned}&\langle\psi_0(\mathbf{x},\tau)\bar{\psi}_0(\mathbf{x}',\tau')\rangle\\ &= \langle(\rho(\mathbf{x},\tau)\rho(\mathbf{x}',\tau'))^{1/2}\exp \mathrm{i}(\varphi(\mathbf{x},\tau)-\varphi(\mathbf{x}',\tau'))\rangle\\ &= \frac{\int D\rho D\varphi(\rho(\mathbf{x},\tau)\rho(\mathbf{x}',\tau'))^{1/2}\exp[S_{\text{eff}}+\mathrm{i}(\varphi(\mathbf{x},\tau)-\varphi(\mathbf{x}',\tau'))]}{\int D\rho D\varphi\exp S_{\text{eff}}}. \quad (7.28)\end{aligned}$$

In the first approximation we may replace $\rho(\mathbf{x},\tau)$, $\rho(\mathbf{x}',\tau')$ by $\rho_0(k_0)$. Moreover, we may only take into account the quadratic form in S_{eff}. Then

$$\begin{aligned}&\langle\psi_0(\mathbf{x},\tau)\bar{\psi}_0(\mathbf{x}',\tau')\rangle \approx \rho_0(k_0)\exp(-\tfrac{1}{2}\langle(\varphi(\mathbf{x},\tau)-{}^2\varphi(\mathbf{x}',\tau'))^2\rangle)\\ &= \rho_0(k_0)\exp\left(-\frac{1}{2\beta V}\sum_{\omega,|\mathbf{k}|<k_0}\frac{m}{\rho(k^2+\omega^2c^{-2})}|e^{\mathrm{i}(\omega\tau+\mathbf{kx})}-e^{\mathrm{i}(\omega\tau'+\mathbf{kx}')}|^2\right). \quad (7.29)\end{aligned}$$

Here we made use of

$$\langle\varphi(p)\varphi(-p)\rangle_0 = \frac{m}{\rho(k^2+\omega^2c^{-2})}. \quad (7.30)$$

This is exactly the element $(G(p))_{11}$ of the matrix $G(p)$ inverse to $G^{-1}(p)$ (7.21), where the term $k^2/4m\rho_0$ is omitted for small k. The value c^2 in (7.30) is the squared sound velocity for which we have

$$c^2 = \frac{p_\lambda}{m}\left(p_{\lambda\lambda}-\frac{p^2_{\lambda\rho_0}}{p_{\rho_0\rho_0}}\right)^{-1} = \frac{1}{m}\frac{\mathrm{d}p}{\mathrm{d}\rho}. \quad (7.31)$$

(The condition $p_{\rho_0}=0$ implies that $p_{\lambda\lambda}-p^2_{\lambda\rho_0}/p_{\rho_0\rho_0}$ equals the full derivative $\mathrm{d}^2p(\lambda,\rho_0(\lambda))/\mathrm{d}\lambda^2$.)

Now we can show that (7.29) behaves like $r^{-\alpha}$ for large $r=|\mathbf{x}-\mathbf{x}'|$ in both cases $d=2$, T small and $d=1$, $T=0$.

Let us begin with the case $d=2$. Putting $\tau'=\tau$ we obtain

$$|e^{\mathrm{i}\mathbf{kx}}-e^{\mathrm{i}\mathbf{kx}'}|^2 = 2(1-\cos(\mathbf{k},\mathbf{x}-\mathbf{x}')).$$

So we come to the following expression for the exponent on the right-hand side of (7.29):

$$-\frac{mc^2}{\beta\rho(2\pi)^2}\int_{k<k_0} d^2k(1-\cos(\mathbf{k},\mathbf{x}-\mathbf{x}'))\sum_{\omega}(\omega^2+c^2k^2)^{-1}$$

$$=-\frac{mc^2}{2\rho(2\pi)^2}\int_{k<k_0}\frac{d^2k}{ck}(1-\cos(\mathbf{k},\mathbf{x}-\mathbf{x}'))\operatorname{cotanh}\tfrac{1}{2}\beta ck$$

$$=-\frac{mc}{4\pi\rho}\int_0^{k_0} dk(1-J_0(kr))\left(1+\frac{2}{e^{\beta ck}-1}\right)$$

$$\approx-\frac{mck_0}{4\pi\rho}-\frac{mc}{2\pi\rho}\int_0^{\infty}\frac{1-J_0(kr)}{e^{\beta ck}-1}dk. \tag{7.32}$$

Here J_0 is the Bessel function. The integral on the right-hand side of (7.32) behaves like

$$\int_0^{\infty}\frac{1-J_0(kr)}{e^{\beta ck}-1}dk=\frac{1}{\beta c}\left[\ln\frac{r}{2\beta c}+C+\cdots\right] \tag{7.33}$$

for large r ($r \gg \beta c$). C in (7.33) is the Euler constant. So we obtain the following result for (7.29):

$$\rho_0(k_0)\exp\left(-\frac{mck_0}{4\pi\rho}\right)\exp\left(-\frac{m}{2\pi\beta\rho}\left(\ln\frac{r}{2\beta c}+C\right)\right), \tag{7.34}$$

which yields the claimed $r^{-\alpha}$ behaviour, where

$$\alpha=\frac{m}{2\pi\beta\rho}. \tag{7.35}$$

The coefficient $\rho_0(k_0)\exp(-mck_0/4\pi\rho)$ must be independent of k_0. It can be shown that this is really the case and that this coefficient is equal to $\rho_0(T=0)$, i.e. the condensate density of the system at $T=0$. Thus we can write down the final result for the two-dimensional case,

$$\langle\psi(\mathbf{x},\tau)\bar{\psi}(\mathbf{x}',\tau')\rangle\approx\rho_0(T=0)\exp\left(-\frac{m}{2\pi\beta\rho}\left(\ln\frac{r}{2\beta c}+C\right)\right). \tag{7.36}$$

It turns out that the $r^{-\alpha}$-like behaviour of the average (7.27) holds for all $T<T_c$, with

$$\alpha=\frac{m}{2\pi\beta\rho_s}, \tag{7.37}$$

where ρ_s is the so-called superfluid density of the system. For small T $\rho_s\approx\rho$ and we come back to (7.35).

It is not difficult to investigate (7.29) also for the case $d = 1$, $T = 0$. The result is

$$\langle \psi(\mathbf{x}, \tau)\bar{\psi}(\mathbf{x}', \tau)\rangle \sim R^{-\gamma}, \tag{7.38}$$

where

$$R^2 = (x - x')^2 + c^2(\tau - \tau')^2, \quad \gamma = \frac{mc}{2\pi\rho}. \tag{7.39}$$

If $d = 1$, and T is arbitrary small, the average (7.29) decreases exponentially like

$$\exp(-(m/2\beta\rho)r). \tag{7.40}$$

Let us emphasize that all the results given by (7.36), (7.38) and (7.40) are obtained from the same formula (7.29).

We see that the method developed in this section allows us to build up the effective action functional and to obtain long-wavelength asymptotics of Green's functions for superfluid systems. The method may also be applied for constructing the hydrodynamical Hamiltonian in superfluidity theory. Landau (1941) postulated the so-called hydrodynamical Hamiltonian

$$\int \mathrm{d}^3x(\tfrac{1}{2}\hat{\mathbf{v}}\hat{\rho}\hat{\mathbf{v}} + F(\hat{\rho})) \tag{7.41}$$

for the description of quantum liquid. Here $\hat{\rho}$ is the density operator, and $\hat{\mathbf{v}}$ the velocity operator; they obey the following commutation relations

$$[\hat{\rho}(\mathbf{x}), \hat{\mathbf{v}}(\mathbf{y})] = (\mathrm{i}/m)\nabla\delta(\mathbf{x} - \mathbf{y}). \tag{7.42}$$

Using this Hamiltonian, Landau could explain hydrodynamical behaviour of superfluid liquid helium.

The theory of Landau is semiphenomenological. The first microscopic model of superfluid (nonideal Bose gas) was considered by Bogoliubov (1947).

The method of integration over fast and slow fields can be used for constructing Landau's hydrodynamical Hamiltonian. In fact, in this section we have obtained not a hydrodynamical Hamiltonian, but the corresponding effective action functional after integration with respect to 'fast' fields. It is in this way that the superfluidity without Bose condensation in two-dimensional and one-dimensional systems was developed (Popov, 1971, 1972*a*, *b*, 1973). Other approaches to the theory of two-dimensional superfluids were suggested by Berezinsky (1971) and also by Kosterlitz & Thouless (1973).

8

Quantum vortices in superfluids

In addition to the acoustic (phonon) excitations described above, specific excitations called 'quantum vortices' may exist in superfluid Bose systems.

Attempts to incorporate the vortex excitations have already been made in the original version of Landau's theory (Landau, 1941). The existence of a periodical lattice of quantum vortices in a superconductor located in a magnetic field was conjectured by Abrikosov (1952, 1957). The idea of quantum vortices occurring in rotating helium below the λ-point was introduced independently by Onsager (1949) and Feynman (1955).

In this section we present the description of quantum vortices in the functional integral formalism. We will also discuss the role of quantum vortices in the phase transition for the superfluid to the normal state.

Our starting point is the functional integration scheme over fast and slow variables. We will apply this formalism mainly to two-dimensional Bose systems and will try to take into account vortex-like configurations of the Bose field. We thus come to the conclusion that the system of vortices is equivalent to the system of charged particles interacting via a new field which play the role of electromagnetic field. At small temperatures there are only pairs with opposite signs, bounded by the long-range logarithm potential. It turns out that the phase transition in the two-dimensional Bose system is nothing but the dissociation of bounded pairs. A similar approach to the three-dimensional systems leads to the conclusion that phase transition is accompanied here by creation of long vortex filaments. Both when $d=2$ and $d=3$ the vortex-like excitations play an important role in the mechanism of the phase transition.

We begin with the formula

$$\int D\bar{\psi}_1 D\psi_1 \exp S[\psi, \bar{\psi}] = \exp S_{\text{eff}}[\psi_0, \bar{\psi}_0], \tag{8.1}$$

where $S[\psi, \bar{\psi}]$ is the starting action functional, $S_{\text{eff}}[\psi_0, \bar{\psi}_0]$ is the effective action of the 'slow' field. For the two-dimensional Bose system S_{eff} is equal to

$$\int \mathrm{d}\tau \mathrm{d}^2x \bigg(ip_{\lambda\rho_0}\pi\partial_\tau\varphi - \frac{p_\lambda}{2m}(\nabla\varphi)^2 - \tfrac{1}{2}p_{\lambda\lambda}(\partial_\tau\varphi)^2 + \tfrac{1}{2}p_{\rho_0\rho_0}\pi^2 - \frac{(\nabla\pi)^2}{8m\rho_0} - \frac{\pi(\nabla\varphi)^2}{2m} \bigg). \tag{8.2}$$

This formula is valid at low temperatures. Let us note that if $T \neq 0$ the coefficient in front of $-(2m)^{-1}(\nabla\varphi)^2$ in (8.2) has the meaning of superfluid density ρ_s. The superfluid density ρ_s is equal to the full density $\rho = p_\lambda$ at zero temperature. Taking this into account, we will replace p_λ by ρ_s, if $T \neq 0$.

When we derived (8.2), we took into account only the field functions $\psi_0(\mathbf{x}, \tau)$, $\bar{\psi}_0(\mathbf{x}, \tau)$, which had no zeros in the $\mathbf{x}, \tau$ space (more precisely $|\psi_0(\mathbf{x}, \tau)|^2 = \rho(\mathbf{x}, \tau)$ was close to $\rho_0(k_0)$). Now let us consider the case when $\psi_0(\mathbf{x}, \tau)$ may vanish at some discrete set of points in the $\mathbf{x}$ plane (for a fixed τ). If we go around such a point, the phase variable $\varphi(\mathbf{x}, \tau)$ acquires an increment $2\pi n$, where n is an integer. Let us consider only the points with $n = \pm 1$. We may say that a point $\mathbf{x}$, where $\psi_0 = \bar{\psi}_0 = 0$, is the center of a quantum vortex rotating around this point.

Now we are going to take into account vortex-like configurations of the $\psi_0, \bar{\psi}_0$ fields. In the first approximation we can neglect the terms $\sim \pi(\nabla\varphi)^2$, $(\nabla\pi)^2$ in (8.2). If we drop these terms, we can integrate over the π-variable and obtain the following action functional of the φ field

$$-\int \mathrm{d}\tau\, \mathrm{d}^2x \frac{\rho_s}{2m}((\nabla\varphi)^2 + c^{-2}(\partial_\tau\varphi)^2), \tag{8.3}$$

where c is the sound velocity. Using new coordinates $x_1 = x, x_2 = y, x_3 = c\tau$, we arrive at the 'relativistic' action

$$-\frac{\rho_s}{2mc}\int \mathrm{d}^3x(\nabla\varphi)^2, \tag{8.4}$$

where $\nabla\varphi$ is the three-dimensional gradient of the φ field.

In the presence of vortices $\varphi(\mathbf{x})$ is not a single-valued function. We can reduce it in a single-valued function by performing the shift transformation

$$\varphi(\mathbf{x}) \to \varphi(\mathbf{x}) + \varphi_0(\mathbf{x}). \tag{8.5}$$

Here $\varphi_0(\mathbf{x})$ is the multivalued part of $\varphi(\mathbf{x}) + \varphi_0(\mathbf{x})$ obeying the Laplace equation

$$\Delta\varphi_0(\mathbf{x}) = 0 \tag{8.6}$$

and the new $\varphi(\mathbf{x})$ is single-valued. The expression (8.4) turns out to be equal to

$$-\frac{\rho_s}{2mc}\int(\nabla\varphi)^2\mathrm{d}^3x-\frac{\rho_s}{2mc}\int(\nabla\varphi_0)^2\mathrm{d}^2x. \tag{8.7}$$

How can one determine $\varphi_0(\mathbf{x})$? It is not difficult to check that $\mathbf{h}(\mathbf{x}) = \nabla\varphi_0(\mathbf{x})$ is nothing but the solution of the following magnetostatic problem in three-dimensional space $x_1, x_2, x_3 = c\tau$:

$$\operatorname{curl}\mathbf{h} = 2\pi\mathbf{j}, \quad \operatorname{div}\mathbf{h} = 0. \tag{8.8}$$

Here $\mathbf{j}(\mathbf{x})$ is the current density corresponding to linear unit currents following along the vortex lines:

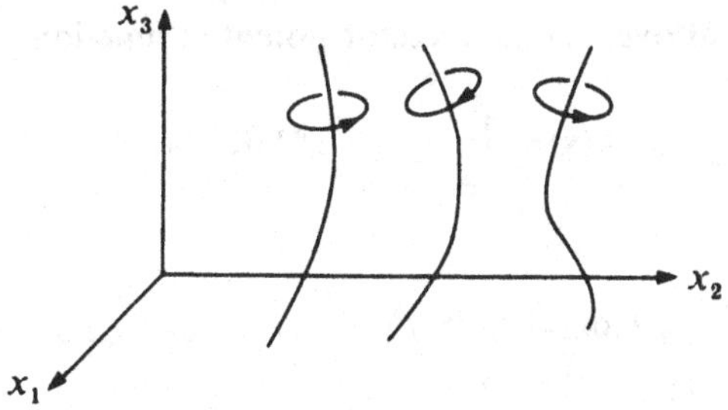

The function $\varphi_0(\mathbf{x})$ is a multi-valued scalar potential of the magnetic field generated by the system of linear currents. The integral

$$\int(\nabla\varphi_0)^2\mathrm{d}^3x = \int h^2\mathrm{d}^3x \tag{8.9}$$

is the energy of the magnetic field.

The magnetostatic problem (8.8) can be solved by means of the vector-potential function $\mathbf{a}(\mathbf{x})$ so that

$$\mathbf{h} = \operatorname{curl}\mathbf{a}, \quad \operatorname{div}\mathbf{a} = 0. \tag{8.10}$$

In the case of the system of linear currents we have

$$\mathbf{a}(\mathbf{x}) = \tfrac{1}{2}\sum_i\int\frac{\mathrm{d}\mathbf{l}_i(\mathbf{y})}{|\mathbf{x}-\mathbf{y}|}. \tag{8.11}$$

This formula leads to

$$\int h^2\mathrm{d}^3x = \pi\sum_{i,k}\int\int\frac{(\mathrm{d}\mathbf{l}_i(\mathbf{x}),\mathrm{d}\mathbf{l}_k(\mathbf{y}))}{|\mathbf{x}-\mathbf{y}|}. \tag{8.12}$$

The integrals with $i = k$ diverge when $\mathbf{x}$ is close to $\mathbf{y}$. This divergence is the result of the approximation in which the vortices are considered as point-like objects. We can take into account the finite size of vortices by substituting

$$\frac{\pi\rho_s}{2mc}\sum_i\int\int_{|\mathbf{x}-\mathbf{y}|<r_0}\frac{(\mathrm{d}\mathbf{l}_i(\mathbf{x}),\mathrm{d}\mathbf{l}_i(\mathbf{y}))}{|\mathbf{x}-\mathbf{y}|}\to\frac{E_v(r_0)}{c}\sum_i\int\mathrm{d}\mathbf{S}_i, \tag{8.13}$$

where $\mathrm{d}\mathbf{S}=|\mathrm{d}\mathbf{l}|=(\mathrm{d}S^2)^{1/2}$. Here $E_v(r_0)$ is the energy of the vortex inside the (circle of) radius r_0. The function $E_v(r_0)$ depends on r_0 logarithmically:

$$E_v(r_0)=\frac{\pi\rho_s}{m}\ln\frac{r_0}{a}, \tag{8.14}$$

where a is the vortex core radius. This formula is valid for $r_0 \gg a$.

We now express the contribution of those x, y for which $|\mathbf{x}-\mathbf{y}| > r_0$ as a functional integral over the new vector potential function

$$\mathbf{A}(\mathbf{x})=\int_{|\mathbf{k}|<k_0\sim r_0^{-1}}\mathrm{e}^{\mathrm{i}\mathbf{k}\mathbf{x}}\mathbf{a}(\mathbf{k})\,\mathrm{d}k. \tag{8.15}$$

We have

$$\exp\left(-\frac{\pi\rho_s}{2mc}\sum_{i,k}\int\int\frac{(\mathrm{d}\mathbf{l}_i(\mathbf{x}),\mathrm{d}\mathbf{l}_k(\mathbf{y}))}{|\mathbf{x}-\mathbf{y}|}\right)$$

$$=\frac{\int\exp\left(-\frac{1}{2c}\int\mathrm{d}^3x((\mathrm{curl}\,\mathbf{A})^2-\mathrm{i}q(\mathbf{A}\mathbf{j}))\right)\prod_x\delta(\partial_iA_i)\prod_i\mathrm{d}A_i}{\int\exp\left(-\frac{1}{2c}\int\mathrm{d}^3x(\mathrm{curl}\,\mathbf{A})^2\right)\prod_x\delta(\partial_iA_i)\prod_i\mathrm{d}A_i}, \tag{8.16}$$

where

$$q=2\pi\left(\frac{\rho_s}{mc^2}\right)^{1/2} \tag{8.17}$$

is a coupling constant. We can prove (8.16) by making a shift transformation which cancels the linear term $\int(\mathbf{j}\mathbf{A})\mathrm{d}^3x$ in the exponent of the integrand in the numerator on the right-hand side of (8.16). We then obtain the following effective action for a Bose system with quantum vortices:

$$S_{\mathrm{eff}}=-m_v(r_0)c\sum_i\int\mathrm{d}S_i-\mathrm{i}q\int(\mathbf{A}\mathbf{j})\mathrm{d}^3x-\frac{1}{2c}\int(\mathrm{curl}\,\mathbf{A})^2\mathrm{d}^3x, \tag{8.18}$$

where

$$m_v(r_0)=E_v(r_0)c^{-2} \tag{8.19}$$

is the vortex mass. The effective action (8.18) describes a system of charged particles interacting via the electromagnetic field. So we have to integrate over the electromagnetic field $\mathbf{A}(x)$ and over the trajectories of charged particles.

If we want to describe the motion of vortices with velocities much smaller than c, we have to consider the nonrelativistic approximation. In this case we have

$$m_v(r_0)c\int dS_i \approx E_v(r_0)\int_0^\beta\left(1+\frac{1}{2c^2}\left(\frac{d\mathbf{x}_i}{d\tau}\right)^2\right)d\tau$$
$$=\beta E_v(r_0)+\int_0^\beta \frac{m_v(r_0)}{2}\mathbf{v}_i^2(\tau)\,d\tau. \tag{8.20}$$

We can then transform the terms in S_{eff} containing A_0 into the potential interaction term. This gives the following expression for the effective section in the nonrelativistic approximation:

$$-\sum_i\left[\beta E_v(r_0)+\int_0^\beta d\tau\left(\tfrac{1}{2}m_v(r_0)\mathbf{v}_i^2(\tau)+\frac{ig_i}{c}(\mathbf{v}_i,\mathbf{A}_i)\right)\right]$$
$$-\frac{1}{2}\int d\tau\,d^2x[(\partial_1 A_2-\partial_2 A_1)^2+c^{-2}(\partial_\tau \mathbf{A})^2]$$
$$+\frac{1}{4\pi}\int d\tau\,d^2x\,d^2y\,j_0(\mathbf{x},\tau)j_0(\mathbf{y},\tau)\ln|\mathbf{x}-\mathbf{y}|. \tag{8.21}$$

Here A_i is the vector potential in the center of the ith vortex,

$$\frac{g^2}{4\pi}=\frac{\pi\rho_s}{m}. \tag{8.22}$$

The function

$$j_0(\mathbf{x},\tau)=\sum_i g_i\delta(\mathbf{x}-\mathbf{x}_i(\tau)) \tag{8.23}$$

is the charge density.

We have shown the equivalence between the Bose system with quantum vortices and the two-dimensional system of charged particles interacting via an electromagnetic field. Now let us consider some consequences of this equivalence.

It is obvious that only pairs of quantum vortices, bound by the long-range logarithmic potential, may be present in the system at low temperatures. As T increases, the number of bound pairs also increases while the average distance between pairs decreases. Supposing that there exists a temperature T_c of dissociation of bound pairs, such that for $T > T_c$ there will be not only bound pairs but single vortices as well. In this case we have a plasma-like state. It can be shown that the correlator $\langle \psi(\mathbf{x}, \tau)\bar{\psi}(\mathbf{x}', \tau')\rangle$ decreases exponentially in such a state due to the Debye screening mechanism (see section 9). Now it is natural to identify this dissociation phase transition with the phase transition from the superfluid to the normal state.

Let us note that the quantity ρ_s, which is the coefficient of $(2m)^{-2}(\nabla\varphi)^2$ in (8.4) does not vanish for $T > T_c$. We can denote this quantity by $\rho_s^{(0)}$ or $\rho_{s(\mathrm{Micro})}$ in order to distinguish it from the macroscopic superfluid density $\rho_{s(\mathrm{Macro})}$, which must disappear for $T > T_c$.

We come to the conclusion that the description of the system in terms of quantum vortices may be valid not only for $T < T_c$ but also for $T > T_c$.

The method outlined here can be applied to three-dimensional superfluid systems. In such systems there are vortex-like excitations which have the form of vortex rings. The number of such rings in volume is small at low temperature but increases when the system is heated. If the average distance between the rings becomes comparable with the average size of the ring, there develops a tendency to create long vortex filaments. We can suppose that the phase transition is accompanied by the appearance of (infinitely) long vortex filaments in the system. Such a qualitative picture leads to the conclusion that all the difference between the superfluid and normal states of a liquid is due to different characters of vortex excitations. In the normal state there exist not only vortex rings but also (infinitely) long vortex filaments.

It is interesting that quantum vortices in rotating liquid helium were observed by E. L. Andronikashvili and others (Andronikashvili & Mamaladze, 1966) approximately 20 min after the system was heated to 0.2 K above the phase transition temperature (λ-point).

The approach outlined above gives a qualitative explanation of this experiment. The existence of quantum vortices for $T > T_c$ is possible because $\rho_{s(\mathrm{Micro})}$ does not vanish. On the other hand, we can explain the fact that the regular structure of vortices breaks down 20 min after heating above the phase transition temperature. The point is that the appearance of long vortex filaments may destroy the regular structure of quantum vortices.

In conclusion, we present a summary of the main result obtained by the functional integration method in the superfluidity theory in sections 5–8.

(1) The perturbation theory of superfluid Bose systems can easily be constructed by means of the functional integration method (section 5).

(2) The application of this perturbative scheme to the low-density Bose gas allows us to obtain Green's functions and thermodynamical functions as well as the phase transition temperature (section 6).

(3) The modification of the method of successive integration over fast and slow variables is suitable for the construction of the effective (hydrodynamical) action functional and the derivation of the low-frequency asymptotic behaviour of Green's functions (section 7).

(4) The method of effective action functional can be extended to two-dimensional and one-dimensional Bose systems, in which the superfluidity may occur without the Bose-condensate. In this case, the phase transition corresponds to emerging of long-range correlations with power-like (and not exponential) asymptotic behaviour (section 7).

(5) In the formalism of functional integration, quantum vortices correspond to the zeros of functions $\psi(\mathbf{x}, \tau), \bar{\psi}(\mathbf{x}, \tau)$, the functional integration variables. In a superfluid state, the main contribution to the functional integral is given by the functions $\psi, \bar{\psi}$ without zeros, which describe the state without quantum vortices. In a normal state, the most important contribution comes from the functions of the type of superposition of plane waves. In the neighbourhood of phase transition, the maximal contribution to the integral comes from the functions whose modulus is almost constant everywhere except the vortex core (see section 8).

(6) Quantum vortices may also occur in a normal Bose system. In a two-dimensional system, the phase transition from the superfluid to the normal state corresponds to the dissociation of coupled pairs; in a three-dimensional system to the emerging of long vortex lines. So, quantum vortices play an important role in the mechanism of phase transition for both two-dimensional and three-dimensional Bose systems.

Part III

Plasma and superfluid Fermi systems

9
Plasma theory

In this section we consider some applications of the functional integral method to the theory of systems with the long-range Coulomb interaction between particles. We are going to demonstrate that the idea of integrating first over the fast variables and then over the slow ones is as useful in plasma physics as in the theory of superfluidity.

We will construct the effective action functional describing the collective behaviour of plasma. As a first example of the application of the method we will obtain the Debye screening. Then we will consider the problem of plasma oscillations and their damping.

For the sake of simplicity we consider a model of electrons in a constant background potential of a charge of opposite sign.

We start with the following action functional:

$$
\begin{aligned}
S = \int_0^\beta \mathrm{d}\tau \int \mathrm{d}^3x(&\bar{\psi}(\mathbf{x},\tau)\partial_\tau\psi(\mathbf{x},\tau) \\
&- (2m)^{-1}\nabla\bar{\psi}(\mathbf{x},\tau)\nabla\psi(\mathbf{x},\tau) + \lambda\bar{\psi}(\mathbf{x},\tau)\psi(\mathbf{x},\tau)) \\
&- \frac{e^2}{2}\int_0^\beta \mathrm{d}\tau \int \frac{\mathrm{d}^3x\mathrm{d}^3y}{|\mathbf{x}-\mathbf{y}|}\bar{\psi}(\mathbf{x},\tau)\bar{\psi}(\mathbf{y},\tau)\psi(\mathbf{y},\tau)\psi(\mathbf{x},\tau).
\end{aligned} \tag{9.1}
$$

We can take into account the background if we drop the term $k = 0$ in the Fourier series for the potential

$$
\frac{4\pi e^2}{V}\sum_{\mathbf{k}\neq 0}\frac{\mathrm{e}^{\mathrm{i}(\mathbf{k},\mathbf{x}-\mathbf{y})}}{k^2}. \tag{9.2}
$$

The standard diagram techniques have the following elements:

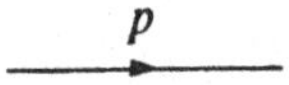

$$
\left(\mathrm{i}\omega - \frac{k^2}{2m} + \lambda\right)^{-1}, \quad \omega = (2n+1)\pi/\beta
$$

$$
4\pi e^2(|\mathbf{k}_1 - \mathbf{k}_3|^2 - |\mathbf{k}_1 - \mathbf{k}_4|^{-2}). \tag{9.3}
$$

The vertex function in (9.3) becomes infinite as $|\mathbf{k}_1 - \mathbf{k}_3| \to 0$ or $|\mathbf{k}_1 - \mathbf{k}_4| \to 0$. Thus if we integrate products of vertex functions, we encounter singularities.

Now we want to build up a new perturbative scheme in order to avoid such singularities. For that purpose we introduce a new field variable φ in addition to $\psi, \bar{\psi}$ fields.

Consider the Gaussian integral over the real Bose field

$$\int \mathrm{D}\varphi \exp\left[-(8\pi)^{-1}\int \mathrm{d}\tau \mathrm{d}^3x (\nabla\varphi(\mathbf{x},\tau))^2\right] \tag{9.4}$$

and insert it into the integral over the $\psi, \bar{\psi}$ variables. Next perform the shift transformation

$$\varphi(\mathbf{x},\tau) \to \varphi(\mathbf{x},\tau) + \mathrm{i}e\int \frac{\mathrm{d}^3y}{|\mathbf{x}-\mathbf{y}|}\bar{\psi}(\mathbf{y},\tau)\psi(\mathbf{y},\tau), \tag{9.5}$$

which cancels the Coulomb interaction term in (9.1).

We thereby obtain the action functional

$$\tilde{S} = \int \mathrm{d}\tau\, \mathrm{d}^3x[\bar{\psi}\partial_\tau\psi - (2m)^{-1}\nabla\bar{\psi}\nabla\psi + \lambda\bar{\psi}\psi - \mathrm{i}e\varphi\bar{\psi}\psi - (8\pi)^{-1}(\nabla\varphi)^2], \tag{9.6}$$

which replaces (9.1). $\tilde{S}$ describes a system of Fermi particles interacting with the field of scalar electric potential.

Perturbation theory for (9.6) has the following elements:

 $\left(\mathrm{i}\omega - \dfrac{k^2}{2m} + \lambda\right)^{-1}$

 $4\pi k^{-2}$ (9.7)

 $\mathrm{i}e.$

Actually this perturbative scheme does not differ from the initial one. It still contains infrared singularities. In order to avoid them, we will apply the idea of integration over the fast and slow variables.

Let us define the fast and slow components of the φ field by

$$\varphi_0(\mathbf{x},\tau) = (\beta V)^{-1/2}\sum_{\omega,|\mathbf{k}|\leqslant k_0} \mathrm{e}^{\mathrm{i}(\omega\tau+\mathbf{k}\mathbf{x})}\varphi(p),$$

$$\varphi_1(\mathbf{x},\tau) = (\beta V)^{-1/2}\sum_{\omega,|\mathbf{k}|> k_0} \mathrm{e}^{\mathrm{i}(\omega\tau+\mathbf{k}\mathbf{x})}\varphi(p). \tag{9.8}$$

There is a definite physical reason for the introduction of the fast and slow φ

fields. The point is that if we want to describe pair collisions of electrons with a small momentum exchange we have to take into account collective effects. On the other hand, if we consider processes with a large momentum exchange, we may neglect the presence of other particles.

Now let us integrate over Fermi fields $\psi, \bar{\psi}$:

$$\int D\bar{\psi} D\psi \exp \tilde{S} = \exp S_{\text{eff}}[\varphi]. \tag{9.9}$$

The integral over the fields $\psi, \bar{\psi}$ is Gaussian. It is formally equal to det M, where

$$M = \partial_\tau + (2m)^{-1}\nabla^2 + \lambda - ie\varphi(\mathbf{x}, \tau) \tag{9.10}$$

(see, for instance, (3.9)). So we come down to the following formula

$$\int D\bar{\psi} D\psi \exp \tilde{S} = \exp(\ln \det M[\varphi]/M[0] - (8\pi)^{-1} \int d\tau d^3x (\nabla\varphi)^2). \tag{9.11}$$

Here det $M[\varphi]$ is regularized by replacing it with det $M[\varphi]/M[0]$. Now let us modify this expression slightly in order to get a better understanding of how it depends on the φ_0 field.

We can write the φ_0 field in the following form:

$$\varphi_0(\mathbf{x}, \tau) = \varphi_0(\mathbf{x}) + \tilde{\varphi}_0(\mathbf{x}, \tau), \tag{9.12}$$

where

$$\tilde{\varphi}_0(\mathbf{x}, \tau) = (\beta V)^{-1/2} \sum_{\substack{\omega \neq 0 \\ |\mathbf{k}| < k_0}} e^{i(\omega\tau + \mathbf{kx})} \varphi(p). \tag{9.13}$$

Here $\varphi_0(\mathbf{x})$ is the zero frequency part of the $\varphi_0(\mathbf{x}, \tau)$ field. Before integrating over $\psi, \bar{\psi}$, it is convenient to perform the following transformation:

$$\psi(\mathbf{x}, \tau) \to \psi(\mathbf{x}, \tau) \exp\left(ie \int^\tau \tilde{\varphi}_0(\mathbf{x}, \tau) d\tau \right),$$

$$\bar{\psi}(\mathbf{x}, \tau) \to \bar{\psi}(\mathbf{x}, \tau) \exp\left(-ie \int^\tau \tilde{\varphi}_0(\mathbf{x}, \tau) d\tau \right), \tag{9.14}$$

where

$$\int^\tau \tilde{\varphi}_0(\mathbf{x}, \tau) d\tau = (\beta V)^{-1/2} \sum_{\substack{|\mathbf{k}| < k_0 \\ \omega \neq 0}} \frac{1}{i\omega} e^{i(\omega\tau + \mathbf{kx})} \varphi(p). \tag{9.15}$$

Obviously,

$$\frac{\partial}{\partial \tau}\int^{\tau} \tilde{\varphi}_0(\mathbf{x},\tau)\,\mathrm{d}\tau = \varphi_0(\mathbf{x},\tau).$$

So we can replace the M operator in (9.10) by the operator of the form:

$$M_1 = \partial_\tau + \frac{1}{2m}\left(\nabla + \mathrm{ie}\int^{\tau} \nabla\varphi_0(\mathbf{x},\tau)\,\mathrm{d}\tau\right)^2 + \lambda - \mathrm{ie}(\varphi_0(\mathbf{x}) + \varphi_1(\mathbf{x},\tau)). \tag{9.16}$$

In the following diagrams the φ_0 field is denoted by a double line, and the φ_1 field by a wavy line. We now use, instead of those in (9.7), the following diagram techniques:

$$\left(\mathrm{i}\omega - \frac{k^2}{2m} + \lambda\right)^{-1}$$

$$4\pi k^{-2}$$

$$4\pi k^{-2}$$

$$\mathrm{ie} \tag{9.17}$$

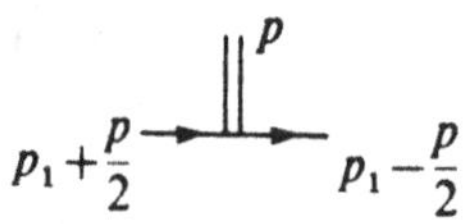

$$\mathrm{ie} \quad \text{if} \quad \omega = 0$$
$$\frac{e(\mathbf{k},\mathbf{k}_1)}{\mathrm{i}m\omega} \quad \text{if} \quad \omega \neq 0$$

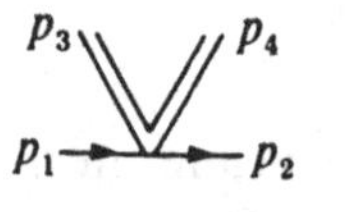

$$-\frac{e^2(\mathbf{k}_3,\mathbf{k}_4)}{m\omega_3\omega_4} \quad \omega_3, \omega_4 \neq 0.$$

Now we may expand the functional $\ln\det M_1[\varphi]/M_1[0]$ as a series in the variables φ_0 and φ_1. Let us write it down graphically

$\ln\det M_1[\varphi]/M_1[0] =$ 

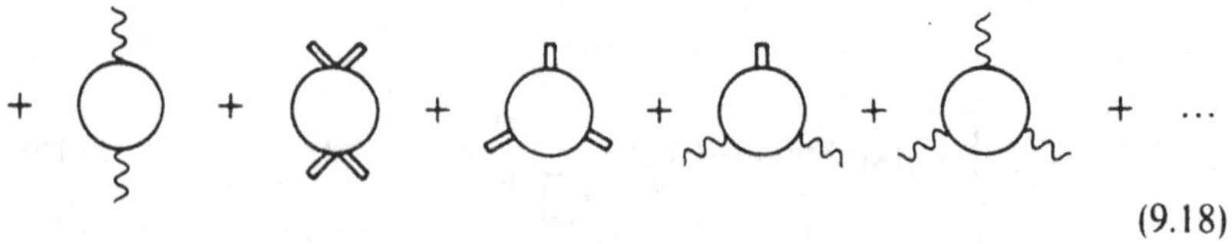

(9.18)

Integrating over the φ_1 field we come to the following result:

$$\int \mathrm{D}\bar{\psi}\mathrm{D}\psi\mathrm{D}\varphi_1 \exp \tilde{S} = \exp S_{\mathrm{eff}}[\varphi_0], \tag{9.19}$$

where

$$S_{\mathrm{eff}}[\varphi_0] = C_0 - (8\pi)^{-1} \int \mathrm{d}\tau\, \mathrm{d}^3x(\nabla\varphi_0(\mathbf{x},\tau))^2$$

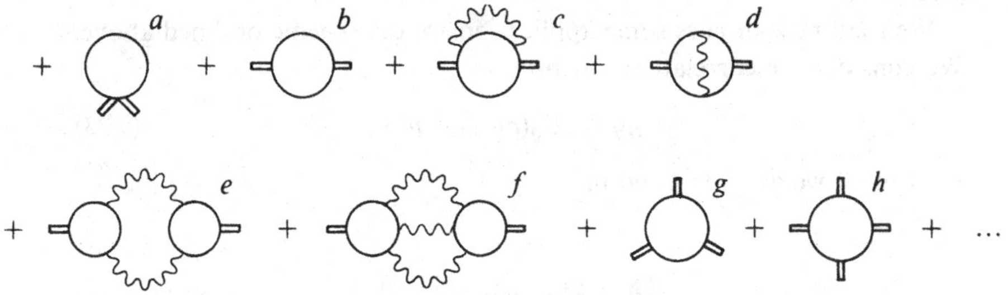

(9.20)

The quadratic form in S_{eff} is due to the integral $(8\pi)^{-1}\int \mathrm{d}\tau\, \mathrm{d}^3x(\nabla\varphi)^2$ and due to all diagrams with two tails $a, b, c, d, e, f, \ldots$. Diagrams g and h are examples of terms of the third and fourth order.

Thus we obtain the following diagram elements for integration over the φ_0 field:

$$G_0(p) = \left(\frac{k^2}{4\pi} + P\right)^{-1}$$

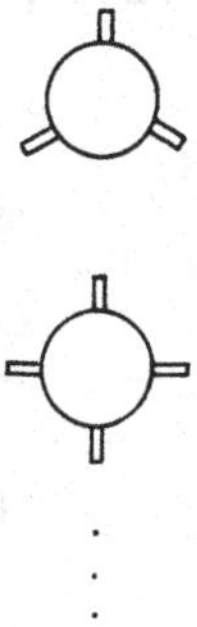

(9.21)

Here

$$P = a + b + c + d + e + f + \cdots \tag{9.22}$$

is the sum of all diagrams with two tails.

The perturbative scheme outlined above has no singularities. The point is that if $\omega = 0$, then $P \neq 0$ and has a nonzero limiting value as $k \to 0$. So $G_0(\omega = 0, \mathbf{k})$ is finite in the limit $k \to 0$ and does not lead to any divergences. If $\omega \neq 0$, $G_0(p)$ is singular because $P \sim k^2$, and $G_0(\omega \neq 0, \mathbf{k}) \sim k^{-2}$. However, the corresponding line in the diagrams goes to two vertices, each of them containing a factor $\mathbf{k}$. This explains why the singularities cancel out. The mechanism of cancellation resembles that in the theory of superfluidity (see section 7).

We shall now discuss some applications of the scheme outlined above. We consider the correlation function

$$G(p) = \langle \varphi_0(p)\varphi_0(-p) \rangle. \tag{9.23}$$

First of all we are interested in

The static case ($\omega = 0$)

For this case it is sufficient to take $G_0(p) = (k^2/4\pi + P)^{-1}$ instead of $G(p)$. Moreover in the first approximation, we may only take into account diagram b in (9.20) (diagram a vanishes if $\omega = 0$). The expression for diagram b is equal to ($\omega = 0$):

$$\begin{aligned} b &= -\frac{e^2}{\beta V} \sum_{\mathbf{k}_1, \omega_1} (\mathrm{i}\omega_1 - \varepsilon(\mathbf{k}_1))^{-1} (\mathrm{i}\omega_1 - \varepsilon(\mathbf{k} + \mathbf{k}_1))^{-1} \\ &= -\frac{e^2}{V} \sum_{k_1} \frac{n(\mathbf{k} + \mathbf{k}_1) - n(\mathbf{k}_1)}{\varepsilon(\mathbf{k} + \mathbf{k}_1) - \varepsilon(\mathbf{k}_1)}. \end{aligned} \tag{9.24}$$

Here $\varepsilon(\mathbf{k}) = k^2/2m - \lambda$, $n(\mathbf{k}) = (\mathrm{e}^{\beta\varepsilon(\mathbf{k})} + 1)^{-1}$. For the Boltzmann plasma $n(\mathbf{k}) \approx \mathrm{e}^{-\beta\varepsilon}$. So for small k we have

$$\frac{n(\mathbf{k} + \mathbf{k}_1) - n(\mathbf{k}_1)}{\varepsilon(\mathbf{k} + \mathbf{k}_1) - \varepsilon(\mathbf{k}_1)} \approx \frac{\partial n(\mathbf{k}_1)}{\partial \varepsilon(\mathbf{k}_1)} = -\beta n(\mathbf{k}_1), \tag{9.25}$$

and we obtain the following result for b:

$$b = \frac{e^2 \beta}{V} \sum_{\mathbf{k}_1} n(\mathbf{k}_1) = e^2 \beta \rho, \tag{9.26}$$

where ρ is the density of the system.

We thus find the following formula for the average (9.23) in the static case:

$$G(0, \mathbf{k}) \approx \frac{4\pi}{k^2 + a^2}, \quad a^2 = 4\pi e^2 \beta \rho. \tag{9.27}$$

The Fourier transform of (9.27)

$$\frac{e^{-ar}}{r} \tag{9.28}$$

is the so-called Debye potential, and $a^{-1}=r_D$ is the radius of Debye screening.

It is clear that if the Debye screening takes place, then a sphere of radius r_D must contain many particles. Then we must have the following inequality:

$$N_D=\frac{4\pi}{3}r_D^3\rho \gg 1. \tag{9.29}$$

Substituting the value of $r_D=a^{-1}$ according to (9.27), we obtain

$$A=T^{3/2}\rho^{-1/2}e^{-3}(3(4\pi)^{1/2})^{-1}\gg 1.$$

So we must have $A^{2/3}\gg 1$, which implies that

$$T\gg(3(4\pi)^{1/2})^{2/3}e^2\rho^{1/3}\sim e^2(\bar{r})^{-1}, \tag{9.30}$$

where $\bar{r}$ is the average distance between two particles. Now T is of the order of the average kinetic energy of a particle and $e^2(\bar{r})^{-1}$ is of the order of the average potential energy. Therefore the condition $N_D\gg 1$ implies that the average potential energy is much smaller than the average kinetic energy. Now we turn to

The dynamical case, $\omega \neq 0$

First of all let us take into account only the diagram a in (9.20). The corresponding expression is equal to ($\varepsilon\to+0$)

$$\begin{aligned} a &= \frac{k^2e^2}{m\omega^2}\frac{1}{\beta V}\sum_{\mathbf{k}_1,\omega_1}\frac{e^{i\omega_1\varepsilon}}{i\omega_1-\varepsilon(\mathbf{k}_1)} \\ &= \frac{k^2e^2}{m\omega^2}\frac{1}{V}\sum_{\mathbf{k}_1}n(\mathbf{k}_1)=\frac{k^2e^2\rho}{m\omega^2}. \end{aligned} \tag{9.31}$$

So we obtain the following formula for $G(p)$:

$$G(p)=\left(\frac{k^2}{4\pi}+P\right)^{-1}\approx\left(\frac{k^2}{4\pi}+\frac{k^2\rho e^2}{m\omega^2}\right)^{-1}=\frac{4\pi}{k^2}\left(1+\frac{4\pi\rho e^2}{m\omega^2}\right)^{-1}. \tag{9.32}$$

Changing $i\omega$ to E, we will find two poles of (9.32) at

$$E=\pm\omega_0,\quad \omega_0^2=\frac{4\pi\rho e^2}{m}; \tag{9.33}$$

ω_0 is the so-called plasma frequency. In the first approximation poles of (9.32) do not depend on **k**.

Now let us also take into account diagram *b* of (9.20). The corresponding expression is equal to

$$-\frac{e^2}{\beta V}\sum_{\mathbf{k}_1,\omega_1}\left(\frac{\mathbf{k}\mathbf{k}_1}{im\omega}\right)^2(\mathrm{i}\omega_1-\varepsilon_1)^{-1}(\mathrm{i}\omega+\mathrm{i}\omega_1-\varepsilon_2)^{-1}$$

$$=\frac{e^2}{V}\sum_{\mathbf{k}_1}\left(\frac{\mathbf{k}\mathbf{k}_1}{m\omega}\right)^2\frac{n_1-n_2}{\mathrm{i}\omega+\varepsilon_1-\varepsilon_2}. \tag{9.34}$$

Replacing $\mathrm{i}\omega$ by E we obtain for small k

$$-\frac{e^2}{V}\sum_{\mathbf{k}_1}\left(\frac{\mathbf{k}\mathbf{k}_1}{mE}\right)^2\frac{n_1-n_2}{E+\varepsilon_1-\varepsilon_2}$$

$$=-\frac{e^2}{V}\sum_{\mathbf{k}_1}\left(\frac{\mathbf{k}\mathbf{k}_1}{mE}\right)^2\frac{(n_1-n_2)(E-\varepsilon_1+\varepsilon_2)}{E^2-(\varepsilon_1-\varepsilon_2)^2}$$

$$=\frac{e^2}{V}\sum_{\mathbf{k}_1}\left(\frac{\mathbf{k}\mathbf{k}_1}{mE}\right)^2\frac{(n_1-n_2)(\varepsilon_1-\varepsilon_2)}{E^2-(\varepsilon_1-\varepsilon_2)^2}$$

$$\approx\frac{e^2}{m^2E^4V}\sum_{\mathbf{k}_1}(\mathbf{k}\mathbf{k}_1)^2(\varepsilon_1-\varepsilon_2)^2\frac{n_1-n_2}{\varepsilon_1-\varepsilon_2}$$

$$\approx\frac{e^2}{m^4E^4V}\sum_{\mathbf{k}_1}(\mathbf{k}\mathbf{k}_1)^4\frac{\partial n}{\partial\varepsilon}=-\frac{3e^2\rho Tk^4}{m^2E^4}. \tag{9.35}$$

The imaginary part of (9.34) is equal to

$$\frac{i\pi e^2}{V}\sum_{\mathbf{k}_1}\left(\frac{\mathbf{k}\mathbf{k}_1}{mE}\right)^2(n_1-n_2)\delta(E+\varepsilon_1-\varepsilon_2)$$

$$=\frac{i\pi e^2}{V}\sum_{\mathbf{k}_1}\left(\frac{\mathbf{k}\mathbf{k}_1}{mE}\right)^2\frac{n_1-n_2}{\varepsilon_1-\varepsilon_2}(\varepsilon_1-\varepsilon_2)\delta(E+\varepsilon_1-\varepsilon_2)$$

$$\approx-\frac{i\pi e^2}{m^2EV}\sum_{\mathbf{k}_1}(\mathbf{k}\mathbf{k}_1)^2\frac{\partial n}{\partial\varepsilon}\delta\left(E-\frac{(\mathbf{k}\mathbf{k}_1)}{m}\right)$$

$$=\mathrm{i}\frac{(2\pi m)^{1/2}\rho\omega_0}{2kT^{3/2}}\exp\left(-\frac{a^2}{2k^2}\right). \tag{9.36}$$

Substituting (9.31) + (9.35) + (9.36) into the Green function we obtain the following result for the pole of $G(p)$:

$$E=E_1(\mathbf{k})=\omega_0+\frac{3Tk^2}{2m\omega_0}-\mathrm{i}\omega_0\left(\frac{\pi}{8}\right)^{1/2}\left(\frac{a}{k}\right)^3\exp\left(-\frac{a^2}{2k^2}\right). \tag{9.37}$$

Here the imaginary part is the so-called Landau damping (Landau, 1946). It decreases faster than any power of k as $k \to 0$. That is why other mechanisms of damping are important for small k.

Besides the Landau damping, there exists the so-called collision damping caused by collisions of charge particles and also a damping caused by plasmon–plasmon interactions. Here we will not go into details but merely summarize the results. One can obtain all these dampings from the effective action S_{eff} (9.20). The result is

$$E(\mathbf{k}) = E_1(\mathbf{k}) - \mathrm{i}\gamma(\mathbf{k}), \tag{9.38}$$

where

$$\gamma(\mathbf{k}) = k^2 \frac{8e^2}{15(\pi m T)^{1/2}} \ln \Lambda \tag{9.39}$$

and

$$\ln \Lambda = \ln \frac{T}{\omega_0} + A \tag{9.40}$$

is the so-called Coulomb logarithm. In (9.40) A is some constant which can be explicitly evaluated as a finite-dimensional integral.

The approach outlined here was developed by V. S. Kapitonov and V. N. Popov (Kapitonov & Popov, 1976). Kapitonov applied these functional methods and derived S_{eff} for the case of plasma in a magnetic field (Kapitonov, 1978).

10

Perturbation theory for superconducting Fermi systems

The effect of superconductivity in Fermi systems is of the same nature as the superfluidity in Bose systems. The main point is again the existence of long-range correlations emerging at the transition temperature. The counterpart of the Bose field in the functional integral formalism is the product of the Fermi fields $\psi_F\psi_F$. Further, the one-particle correlation function of Bose systems

$$\langle \psi(\mathbf{x}, \tau)\bar{\psi}(\mathbf{y}, \tau_1)\rangle \tag{10.1}$$

corresponds to the two-particle correlation function of a Fermi system (i.e. the mean value of the product of four Fermi fields)

$$\langle \psi(\mathbf{x}_1, \tau_1)\psi(\mathbf{x}_2, \tau_2)\bar{\psi}(\mathbf{x}_3, \tau_3)\bar{\psi}(\mathbf{x}_4, \tau_4)\rangle. \tag{10.2}$$

We say that the long-range correlations exist in the system provided correlation function (10.2) goes to zero as an inverse power of $|\mathbf{x}_1 - \mathbf{x}_3|$ or if it has a finite limit as $|\mathbf{x}_1 - \mathbf{x}_3| \to \infty$ while the differences $\mathbf{x}_1 - \mathbf{x}_2$, $\mathbf{x}_3 - \mathbf{x}_4$ (and quantities τ_i) are held fixed. If the limit is finite, one can suppose that there exist anomalous Green's functions

$$\langle \psi(\mathbf{x}_1, \tau_1)\psi(\mathbf{x}_2, \tau_2)\rangle, \langle \bar{\psi}(\mathbf{x}_3, \tau_3)\bar{\psi}(\mathbf{x}_4, \tau_4)\rangle. \tag{10.3}$$

Let us consider the perturbation theory of superconducting Fermi systems supposing that anomalous mean values (10.3) exist. Such mean values are identically zero when calculated in the framework of the ordinary perturbation theory.

One of the possibilities of obtaining a nonzero anomalous mean value is to add to the Fermi system action the 'source functional'

$$\sum_p (\bar{\eta}(p)a(p)a(-p) + \eta(p)a^*(p)a^*(-p)), \tag{10.4}$$

where the complex functions $\eta(p)$, $\bar{\eta}(p)$, $p = (\mathbf{k}, \omega)$ are infinitesimally small. Expression (10.4) does not satisfy the particle number conservation law, and if η and $\bar{\eta}$ are nonzero, then the anomalous mean values (10.3) will also be nonzero. The question is what happens in the limit $\eta, \bar{\eta} \to 0$? If the limit of the anomalous correlation functions is nonzero as $\eta, \bar{\eta} \to 0$, then the system is in superconductive state.

The method of 'source' had been used by Bogoliubov in his formalism of 'quasiaverages' (Bogoliubov, 1961) and since then it has been applied in a lot of papers.

The same result can be obtained using a modification of the skeleton diagram technique, which differs from the skeleton technique introduced in section 2 in the fact that anomalous Green's functions also occur, besides the normal ones. Recall that in the diagram technique the diagram lines correspond to the full Green's functions, but, on the other hand, only diagrams with no subdiagrams of the self-energy type are taken into account. The diagram vertices are identical with those of the ordinary diagram technique.

To give the diagram technique a concrete meaning, one has to consider the Fermi particle spins and describe particles with different spins s by different functions $\psi_s(\mathbf{x}, \tau)$.

Consider the situation when the anomalous mean values of fields with opposite spins

$$\langle \psi_s(\mathbf{x}, \tau)\psi_{-s}(\mathbf{y}, \tau_1)\rangle, \langle \bar{\psi}_s(\mathbf{x}, \tau)\bar{\psi}_{-s}(\mathbf{y}, \tau_1)\rangle \tag{10.5}$$

are nonzero, while the mean values $\langle \psi_s \psi_s \rangle$, $\langle \bar{\psi}_s \bar{\psi}_s \rangle$ identically vanish.

In this case, the diagram technique contains the following elements and corresponding terms:

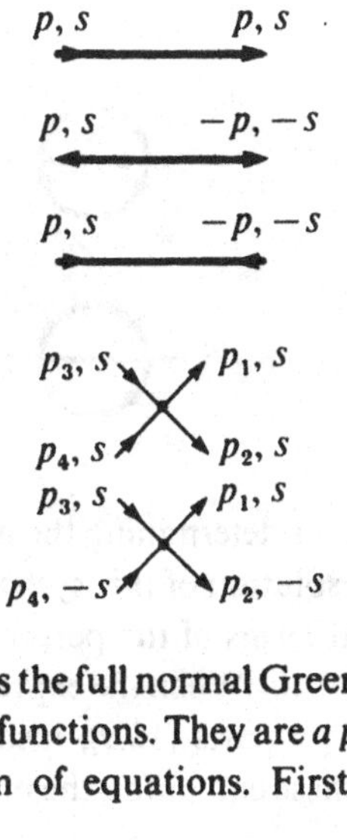

$G(p, s)$

$G_1(p, s)$

$\bar{G}_1(p, s)$ (10.6)

$v(\mathbf{k}_1 - \mathbf{k}_3) - v(\mathbf{k}_1 - \mathbf{k}_4)$

$v(\mathbf{k}_1 - \mathbf{k}_3)$.

Here G is the full normal Green's function and $G_1, \bar{G}_1$ are the full anomalous Green's functions. They are *a priori* unknown and must be found by solving a system of equations. First, this system includes the Dyson–Gorkov equations which express Green's functions in terms of normal $A(p)$ and anomalous $B(p)$ self-energy parts. Graphically, they have the form

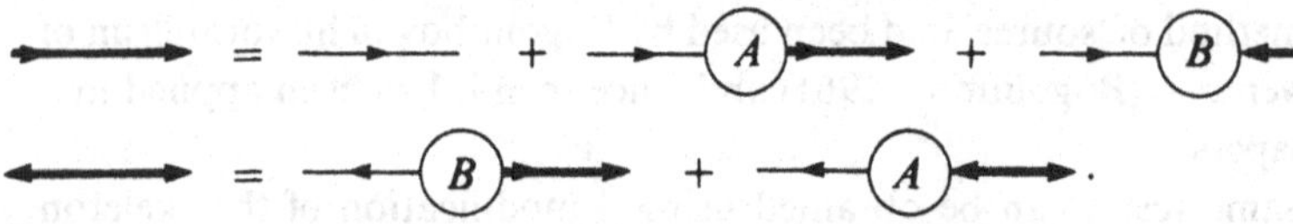

(10.7)

identical to the Dyson–Beliaev equations describing superfluid Bose systems (Beliaev, 1958*a*, *b*). In an analytic form, the equations can be written as

$$G(p) = G_0(p) + G_0(p)A(p)G(p) + G_0(p)B(p)G_1(p),$$
$$G_1(p) = -G_0(-p)\bar{B}(p)G(p) + G_0(-p)A(-p)G_1(p). \tag{10.8}$$

Taking into account that $G_0(p)$ is given by (2.15), we can write the solution of this system in the form

$$G(p) = \left(i\omega + \frac{k^2}{2m} - \lambda + A(-p)\right)Z^{-1}(p),$$

$$G_1(p) = \bar{B}(p)Z^{-1}(p),$$

$$Z(p) = \left(i\omega + \frac{k^2}{2m} - \lambda + A(-p)\right)\left(i\omega - \frac{k^2}{2m} + \lambda - A(p)\right) - |B(p)|^2. \tag{10.9}$$

Each of the self-energy parts $A(p)$, $B(p)$ is given by the sum of the skeleton perturbation theory diagrams

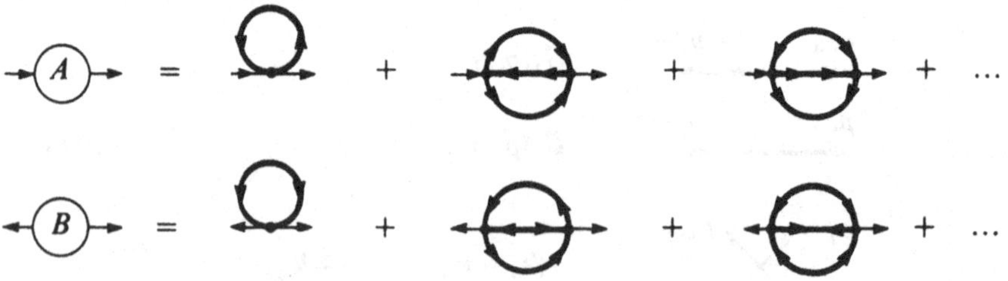

(10.10)

Equations (10.7) and (10.10) form the system for determining the normal and anomalous Green's functions. To find the solution of this system, one has either to restrict oneself to the first several terms of the perturbation series, or find the leading diagram sequence of (10.10). When the expressions of Green's functions in terms of the self-energy parts (10.9), instead of Green's functions themselves, are inserted into these equations, the problem is transformed into the solution of a system of integral equations for self-energy parts.

Let us look for the solution of the system of (10.7) and (10.10) for the model of Fermi gas. Assuming that the gas parameter $a\rho^{1/3}$ is small,

$$a\rho^{1/3} \ll 1, \tag{10.11}$$

it is not difficult to separate the leading sequence of diagrams for A and B. Here, the same factors decreasing the diagram order are in operation as in the Bose gas theory (section 6), namely the anomalous lines and the closed loops of normal lines that can be passed around in the direction of diagram arrows. Thus, the main contribution to the self-energy parts is given by diagrams in which the number of these factors is minimal. They are the diagrams without anomalous lines and with one closed loop of normal lines in the expressions of $A(p)$, and the diagrams without closed loops and with one anomalous line when calculating $B(p)$ (only one diagram of the last type exists, namely the first in the diagrammatic equation (10.10)). The remaining diagrams give contributions containing higher degrees of the gas parameter than those considered above. Thus, we obtain the approximate equations

(10.12)

The four-point function standing in the $A(p)$ diagram represents the diagram sequence

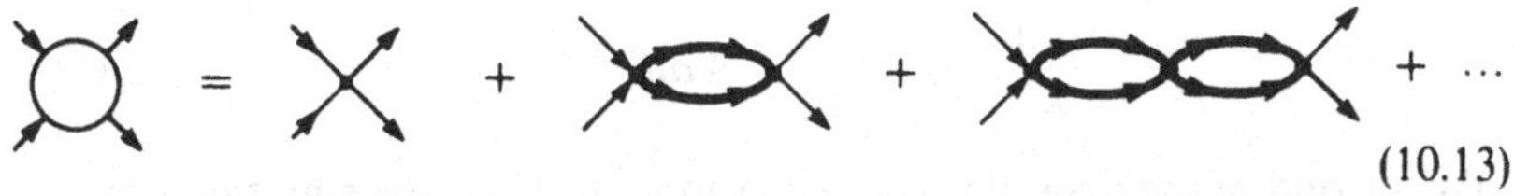

(10.13)

It can be shown in the same way as in the Bose gas theory that the summation of this series results in substituting the potential by the t-matrix which can be put equal to a constant $t_0 = t(\mathbf{k}_1 = 0, \mathbf{k}_2 = 0, z = 0)$ in the domain $k \leqslant k_F = (2m\lambda)^{1/2}$. The vertex describing the scattering of identical spin particles in (10.6) contains the antisymmetrized potential $v(\mathbf{k}_1 - \mathbf{k}_3) - v(\mathbf{k}_1 - \mathbf{k}_4)$. After the summation of (10.13) this expression transforms into the antisymmetrized t-matrix vanishing after the substitution $t \to t_0$. This is the reason why only the t-matrix describing the scattering of opposite spin particles gives a nonzero contribution in the first approximation. For the self-energy $A(p)$ we obtain

$$A_+ = t_0\rho_-, \quad A_- = t_0\rho_+. \tag{10.14}$$

Here $A_\pm$ are the self-energies of particles with spins $\pm$ and $\rho_\pm$ are the corresponding densities. If magnetic field and other forces with an opposite influence on particles with opposite spins are not present, then $A_+ = A_-$, and $\rho_+ = \rho_-$. The densities $\rho_\pm$ can be in the first approximation expressed in terms of the nonperturbative Green's functions $G_0(p)$ ($\varepsilon \to +0$)

$$\rho_+ = \rho_- = (\beta V)^{-1} \sum_p e^{i\omega\varepsilon} G_0(p)$$
$$= (2\pi)^{-3} \int d^3k \left(\exp \beta\left(\frac{k^2}{2m} - \lambda\right) + 1 \right)^{-1} \tag{10.15}$$

In the following we assume that the Fermi gas is degenerated, i.e. that $\beta\lambda \gg 1$. We shall show later that this holds for all temperatures less than the superconducting state transition temperature T_c. In a degenerated Fermi gas the density is determined by the formula

$$\rho_\pm = (2\pi)^{-3} \int_{k<k_F} d^3k = \frac{(2m\lambda)^{3/2}}{6\pi^2}. \tag{10.16}$$

According to (10.14) the self-energy part $A(p)$ is equal to the constant

$$A_\pm = \frac{t_0(2m\lambda)^{3/2}}{6\pi^2}, \tag{10.17}$$

which can be regarded as a small correction $\Delta\lambda$ to the chemical potential λ. Indeed,

$$\frac{|\Delta\lambda|}{\lambda} \sim t_0 m^{3/2} \lambda^{1/2} \sim a\rho^{1/3} \ll 1. \tag{10.18}$$

The second of the diagrammatic equations (10.12) is the equation of the anomalous self-energy part B, which in this approximation does not depend on ω. The anomalous Green's function is in this approximation given by

$$G_1(p) = \frac{B(\mathbf{k})}{\omega^2 + ((k^2/2m) - \mu)^2 + B^2(\mathbf{k})}, \tag{10.19}$$

where the parameter

$$\mu = \lambda - A_\pm = \lambda - \frac{t_0(2m\lambda)^{3/2}}{6\pi^2} \tag{10.20}$$

is the renormalized chemical potential.

The equation for the function B can be given the form

$$B(\mathbf{k}) = -\frac{1}{\beta V}\sum_{p_1} v(\mathbf{k}-\mathbf{k}_1)G_1(p_1)$$

$$= -(2\pi)^{-3}\int d^3k_1 v(\mathbf{k}-\mathbf{k}_1)B(\mathbf{k}_1)\frac{\tanh \beta\varepsilon(\mathbf{k}_1)/2}{2\varepsilon(\mathbf{k}_1)}, \qquad (10.21)$$

where

$$\varepsilon(\mathbf{k}) = \left(\left(\frac{k^2}{2m}-\mu\right)^2 + B^2(\mathbf{k})\right)^{1/2} \qquad (10.22)$$

In order to solve (10.21) we add an identical expression to both sides of this equation and rewrite (10.21) as

$$B(\mathbf{k}) + (2\pi)^{-3}\int v(\mathbf{k}-\mathbf{k}_1)B(\mathbf{k}_1)\frac{m}{k_1^2}d^3k_1$$

$$= (2\pi)^{-3}\int v(\mathbf{k}-\mathbf{k}_1)B(\mathbf{k}_1)\left(\frac{m}{k_1^2} - \frac{\tanh \beta\varepsilon(\mathbf{k}_1)/2}{2\varepsilon(\mathbf{k}_1)}\right)d^3k_1. \qquad (10.23)$$

The operator acting on $B(\mathbf{k})$ on the left-hand side of this equation can be inverted. This is equivalent to replacing the potential $v(\mathbf{k}-\mathbf{k}_1)$ by the t-matrix on the right-hand side. In the domain $k \lesssim k_F$ (which gives the main contribution to the integral) the t-matrix can be replaced by a constant t_0. We obtain the following equation for $B(\mathbf{k})$:

$$B(\mathbf{k}) = \frac{t_0}{(2\pi)^3}\int B(\mathbf{k}_1)\left(\frac{m}{k_1^2} - \frac{\tanh \beta\varepsilon(\mathbf{k}_1)/2}{2\varepsilon(\mathbf{k}_1)}\right)d^3k_1. \qquad (10.24)$$

The equation has a trivial solution $B(\mathbf{k}) = 0$; it may also have a nontrivial solution, corresponding to the superconducting state. Due to (10.24) the function $B(\mathbf{k})$ is constant in the domain $k \lesssim k_F$. This constant can be interpreted as the gap in the energy spectrum obtained from the condition that the denominator of Green's function (10.9) should vanish as $\lambda - A \to \mu$, $B \to \Delta$ after the analytic continuation $i\omega \to E$. The poles of the Green's functions are at $E = \pm\varepsilon(\mathbf{k})$, where $\varepsilon(\mathbf{k})$ is given by formula (10.22), in which $B(\mathbf{k})$ can be substituted by Δ. The quantity Δ is the minimum value of $\varepsilon(\mathbf{k})$ acquired on the Fermi level if $k^2/2m = \mu$. Assuming that $\Delta \neq 0$, we obtain the equation for Δ

$$1 = \frac{mt_0}{2\pi^2}\int_0^\infty dk\left(1 - \frac{k^2}{2m(\xi^2+\Delta^2)^{1/2}}\tanh\frac{\beta}{2}(\xi^2+\Delta^2)^{1/2}\right), \qquad (10.25)$$

where

$$\xi(\mathbf{k}) = \frac{k^2}{2m} - \mu. \tag{10.26}$$

As in the following Δ is shown to be small ($\Delta \ll \lambda$), one can neglect its contribution and replace $\tanh(\beta/2)(\xi^2+\Delta^2)^{1/2}$ by unity everywhere except in a small interval $[k_F - k_0, k_F + k_0]$ close to the Fermi level, where k_0 is chosen such that the inequalities

$$\Delta \ll \frac{k_0^2}{2m} \ll \lambda, \quad \frac{k_0 k_F}{m} \gg T \tag{10.27}$$

hold. Finally we obtain the expressions

$$\int_{|k-k_F|>k_0} \left(1 - \frac{k^2}{2m(\xi^2+\Delta^2)^{1/2}} \tanh\frac{\beta}{2}(\xi^2+\Delta^2)^{1/2}\right) dk$$

$$\approx \int_{|k-k_F|>k_0} \left(1 - \frac{k^2}{|k^2 - k_F^2|}\right) dk = k_F\left(2 - \ln\frac{2k_F}{k_0}\right) - 2k_0,$$

$$\int_{|k-k_F|<k_0} \left(1 - \frac{k^2}{2m(\xi^2+\Delta^2)^{1/2}} \tanh\frac{\beta}{2}(\xi^2+\Delta^2)^{1/2}\right) dk$$

$$\approx 2k_0 - k_F \int_0^{k_F k_0/m} \frac{\tanh(\beta/2)(\xi^2+\Delta^2)^{1/2}}{(\xi^2+\Delta^2)^{1/2}} d\xi$$

for the integrals over $|k-k_F| > k_0$ and $|k-k_F| < k_0$, respectively. Equation (10.25) can be rewritten in the form

$$-1 = \frac{m t_0 k_F}{2\pi^2}\left(-2 + \ln\frac{2k_F}{k_0} + \int_0^{k_F k_0/m} \frac{\tanh(\beta/2)(\xi^2+\Delta^2)^{1/2}}{(\xi^2+\Delta^2)^{1/2}} d\xi\right). \tag{10.28}$$

The solution of this equation exists if t_0 is negative. This means that the interaction of Fermi particles must be attractive. Let

$$\zeta = \frac{m|t_0|k_F}{2\pi^2}; \tag{10.29}$$

The parameter ζ is small ($\zeta \ll 1$) due to the small densities (10.11).

Equation (10.28) only has a solution for sufficiently small temperatures $T = \beta^{-1}$. The transition temperature T_c can be defined as the temperature at which the solution vanishes ($\Delta = 0$). For $\Delta = 0$ we obtain

$$\int_0^{k_0k_F/m} \frac{\tanh(\beta/2)(\xi^2+\Delta^2)^{1/2}}{(\xi^2+\Delta^2)^{1/2}}\mathrm{d}\xi = \int_0^{k_0k_F/2mT}\frac{\tanh x}{x}\mathrm{d}x$$

$$\approx \ln\frac{k_0k_F}{2mT} - \int_0^\infty \frac{\ln x}{\cosh^2 x}\mathrm{d}x = \ln\frac{2k_0k_F\gamma}{\pi mT}, \tag{10.30}$$

where $\ln\gamma = C$ is the Euler constant. The substitution of (10.30) into (20.28) yields

$$T_c = \frac{4\gamma k_F^2}{\pi m}\mathrm{e}^{-\zeta^{-1}-2}. \tag{10.31}$$

We can now find the value $\Delta(0)$ of the gap at $T = 0$. Inserting

$$\int_0^{k_0k_F/m} \frac{\mathrm{d}\xi}{(\xi^2+\Delta^2)^{1/2}} = \ln\frac{2k_0k_F}{m\Delta} \tag{10.32}$$

into (10.28), we obtain

$$\Delta(0) = \frac{4k_F^2}{m}\mathrm{e}^{-\zeta^{-1}-2}. \tag{10.33}$$

The final results, (10.31) and (10.33), do not depend on the auxiliary parameter k_0, as we have supposed. Dividing (10.31) by (10.33), we obtain the relation

$$T_c = (\gamma/\pi)\Delta(0), \tag{10.34}$$

which represents a general property of superconducting theories. It is also valid in the BCS model and in the electron–phonon interaction model.

Finally we notice that the Fermi gas remains degenerated until $T = T_c$ as we have supposed earlier. Indeed

$$T_c \sim \frac{k_F^2}{2m}\mathrm{e}^{-\zeta^{-1}} = \lambda\,\mathrm{e}^{-\zeta^{-1}} \ll \lambda, \tag{10.35}$$

and the degeneracy condition $\beta\lambda \gg 1$ is satisfied for all $T \leqslant T_c$.

11
Superconductivity of the second kind

The interaction of a superconductive system with electromagnetic field results in some interesting effects. One of the most remarkable is the Meissner effect, the consequence of which is that a sufficiently weak magnetic field is pushed out of the superconductor. There are two possible scenarios when the magnetic field is increased:

(1) The field penetrates the superconductor, changing it into the ordinary state. Such superconductors are called 'superconductors of the first kind'.

(2) What emerges in the system is the so-called 'mixed state'. During this process the field partially penetrates the superconductor in which a periodic lattice of quantum vortices is formed. This effect is called 'superconductivity of the second kind', and the superconductors exhibiting this type of superconductivity are called the second kind superconductors.

The creation of the quantum vortex lattice was predicted by Abrikosov (1952) in the framework of the phenomenological superconductivity theory of Ginsburg & Landau (1950). Formulation of the microscopic superconductivity theory allowed the derivation of the Ginsburg–Landau equations microscopically and the understanding of the meaning of the parameters appearing in the equations. In particular, it turned out that the effective charge of the phenomenological theory is equal to twice the electron charge, i.e. that it corresponds to the charge of a Cooper pair.

In this section we shall apply the functional integration method to some of the problems of superconductivity of the second kind. First, we shall show that the homogeneous superconducting state cannot exist in a constant homogeneous magnetic field. Three possibilities are consistent with this fact:

(1) the magnetic field is pushed out of the superconductor (the Meissner effect);

(2) the magnetic field destroys the superconductivity;

(3) the inhomogeneous superconducting state is created.

The last possibility occurs in the second kind superconductors, where the Abrikosov lattice is created. We shall show that the area of the lattice plaquette can be determined purely kinematically, from the relations of gradient invariance.

Further, we shall derive the Ginzburg–Landau equations for a Fermi gas model by means of subsequent functional integration over fast and slow fields. These equations are nothing but the extremality conditions for the functional that has the meaning of effective action of a superconductor. It is well known that the Ginsburg–Landau equations are applicable near the phase transition point where the energy gap in the Fermi excitation spectrum Δ is small in comparison with the value of this gap at $T = 0$. For small temperatures and a weak electromagnetic field one can derive an equation for the function describing the deviation of the ordering parameter from the equilibrium value.

The action functional S of a Fermi system in the electromagnetic field differs from functional (2.3) by the transformation

$$\partial_\tau \to \partial_\tau + ie\varphi, \quad \nabla \to \nabla + ie\mathbf{A}, \tag{11.1}$$

and also by an additional term describing the interaction between spin and the magnetic field. For constant magnetic field $\mathbf{H}$ we have

$$\varphi = 0, \quad \mathbf{A} = \tfrac{1}{2}[\mathbf{H}, \mathbf{x}] \tag{11.2}$$

$$S = \int d\tau\, d\mathbf{x} \sum_s \left(\bar{\psi}_s \partial_\tau \psi_s - (2m)^{-1} |(\nabla + ie\mathbf{A})\psi_s|^2 + \lambda \bar{\psi}_s \psi_s + \frac{eH}{2m} s \bar{\psi}_s \psi_s \right)$$
$$- \frac{1}{2} \int d\tau\, d\mathbf{x}\, d\mathbf{y}\, u(\mathbf{x} - \mathbf{y}) \sum_{s,s'} \rho_s(\mathbf{x}, \tau) \rho_{s'}(\mathbf{y}, \tau), \tag{11.3}$$

where

$$\rho_s(\mathbf{x}, \tau) = \bar{\psi}_s(\mathbf{x}, \tau) \psi_s(\mathbf{x}, \tau) \tag{11.4}$$

is the density of electrons with the helicity s.

Let us now examine the changes of normal and anomalous mean values

$$G(\mathbf{x}, \tau; \mathbf{y}, \tau') = \langle \psi_s(\mathbf{x}, \tau) \bar{\psi}_s(\mathbf{y}, \tau_1) \rangle,$$
$$G_1(\mathbf{x}, \tau; \mathbf{y}, \tau') = \langle \psi_s(\mathbf{x}, \tau) \psi_{-s}(\mathbf{y}, \tau_1) \rangle \tag{11.5}$$

under the translation $\mathbf{x} \to \mathbf{x} + \mathbf{a}$, $\mathbf{y} \to \mathbf{y} + \mathbf{a}$. The mean values (11.5) can be written in terms of functional integrals of the products of averaged anticommuting fields weighted by e^S. We make the substitution $\mathbf{x} \to \mathbf{x} + \mathbf{a}$, $\mathbf{y} \to \mathbf{y} + \mathbf{a}$ in the action S. This yields

$$\mathbf{A} = \tfrac{1}{2}[\mathbf{H}, \mathbf{x}] \to \mathbf{A}' = \tfrac{1}{2}[\mathbf{H}, \mathbf{x}] + \tfrac{1}{2}[\mathbf{H}, a]. \tag{11.6}$$

The functions $\psi_s(\mathbf{x} + \mathbf{a}, \tau)$, $\bar{\psi}_s(\mathbf{x} + \mathbf{a}, \tau)$ will be considered as new functional integration variables. This leads to the conclusion that the mean values (11.5) of fields with shifted arguments are equal to the mean values of fields with original arguments but evaluated with the vector potential $\mathbf{A}'$ in (11.6) instead of $\mathbf{A}$ in (11.2). One can return to the original vector potential $\mathbf{A}$ by phase transformation of the Fermi fields:

$$\psi_s(\mathbf{x},\tau)\to\exp\left(-\frac{ie}{2}([\mathbf{H},\mathbf{a}],\mathbf{x})\right)\psi_s(\mathbf{x},\tau),$$

$$\bar{\psi}_s(\mathbf{x},\tau)\to\exp\left(\frac{ie}{2}([\mathbf{H},\mathbf{a}],\mathbf{x})\right)\bar{\psi}_s(\mathbf{x},\tau). \tag{11.7}$$

We then have the following formulae:

$$G(\mathbf{x}+\mathbf{a},\tau;\mathbf{y}+\mathbf{a},\tau')=\exp\left(-\frac{ie}{2}([\mathbf{H},\mathbf{a}],\mathbf{x}-\mathbf{y})\right)G(\mathbf{x},\tau;\mathbf{y},\tau_1),$$

$$G_1(\mathbf{x}+\mathbf{a},\tau;\mathbf{y}+\mathbf{a},\tau')=\exp\left(-\frac{ie}{2}([\mathbf{H},\mathbf{a}],\mathbf{x}+\mathbf{y})\right)\times G_1(\mathbf{x},\tau;\mathbf{y},\tau_1). \tag{11.8}$$

The second of these formulae indicates that it is impossible to form a homogeneous superconducting state in a homogeneous magnetic field. Indeed, if the superconducting state is homogeneous, then (11.8) hold for any $\mathbf{a}$. Shifting once again $\mathbf{x}\to\mathbf{x}+\mathbf{b}$, $\mathbf{y}\to\mathbf{y}+\mathbf{b}$ in (11.8), we find that the function G_1 acquires the factor

$$\exp(-ie([\mathbf{H},\mathbf{a}],\mathbf{b}))\exp\left(-\frac{ie}{2}([\mathbf{H},\mathbf{a}+\mathbf{b}],\mathbf{x}+\mathbf{y})\right) \tag{11.9}$$

after the translation of its arguments $\mathbf{x}$ and $\mathbf{y}$ by $\mathbf{a}+\mathbf{b}$. If, on the other hand, we simply transform $\mathbf{a}\to\mathbf{a}+\mathbf{b}$ in (11.8), we obtain the factor

$$\exp\left(-\frac{ie}{2}([\mathbf{H},\mathbf{a}+\mathbf{b}],\mathbf{x}+\mathbf{y})\right) \tag{11.10}$$

instead of (11.9). Expression (11.9) can be equal to (11.10) only if the condition

$$\exp ie(\mathbf{H},[\mathbf{a},\mathbf{b}])=1 \tag{11.11}$$

is satisfied. But this cannot be true for arbitrary $\mathbf{a}$ and $\mathbf{b}$. Thus, the homogeneous superconducting state in the magnetic field is impossible.

Equation (11.11) can be satisfied provided the vectors $\mathbf{a}$ and $\mathbf{b}$ form a periodic lattice with the minimal plaguette area

$$\frac{2\pi n}{eH} \tag{11.12}$$

in the plane perpendicular to **H** (n is some positive integer). Putting $n = 1$ and assuming that the lattice is triangular, we obtain the formula

$$a = \left(\frac{4\pi n}{eH3^{1/2}}\right)^{1/2} \tag{11.13}$$

for the lattice spacing a.

Thus we come to the possibility of the existence of a periodic structure in a superconductor.

Let us now derive the effective action functional for the Fermi gas model described by the action (11.3). The extremality conditions of this functional yield equations describing the periodic structure in detail. We apply the idea of successive integration over fast and slow variables to the Fermi systems. We shall call 'the slowly oscillating' part $\psi_{s0}(\mathbf{x}, \tau)$ of the Fermi field $\psi_s(\mathbf{x}, \tau)$ the sum

$$\psi_{s0}(\mathbf{x}, \tau) = (\beta V)^{-1/2} \sum_{\omega, |k - k_F| < k_0} e^{i(\omega\tau + \mathbf{kx})} a_s(p), \tag{11.14}$$

where the momenta $\mathbf{k}$ lie in the narrow layer near the Fermi sphere $|k - k_F| < k_0$. Consequently, the difference $\psi - \psi_0$ will be called 'the fast oscillating part ψ_{1s} of the field ψ_s'.

Carrying out the functional integration over fast variables

$$\int D\bar{\psi}_1 D\psi_1 \exp S = \exp \tilde{S}[\psi_0, \bar{\psi}_0], \tag{11.15}$$

we obtain a functional depending on the slow variables $\psi_0, \bar{\psi}_0$. It is not difficult to calculate the functional $\tilde{S}$ in the case of a low-density Fermi gas. In the first approximation we can restrict ourselves of the terms of an order not higher than the fourth. In this approximation $\tilde{S}$ has the form

$$\tilde{S} = p_0\beta V + \int d\tau d\mathbf{x} \sum_s \Big(\bar{\psi}_{0s}\partial_\tau\psi_{0s} - (2m)^{-1}|(\nabla + ie\mathbf{A})\psi_{0s}|^2$$
$$+ \left(\mu + ie\varphi + \frac{seH}{2m}\right)\bar{\psi}_{0s}\psi_{0s}\Big) - \frac{t'}{2}\int d\tau d\mathbf{x} \bar{\psi}_{0+}\bar{\psi}_{0-}\psi_{0-}\psi_{0+}. \tag{11.16}$$

As compared with the original expression (11.3), the free term $p_0\beta V$ emerges in (11.16), and the chemical potential λ as well as the interaction potential $u(\mathbf{x} - \mathbf{y})$ are replaced by corresponding renormalized quantities μ

and $t'\delta(\mathbf{x}-\mathbf{y})$. The pressure p_0 of nonideal Fermi gas containing no particles with $|k-k_F|<k_0$ is given by the formula

$$p_0 = \frac{T}{(2\pi)^3}\int_{|k-k_F|>k_0} d^3k \ln(1+e^{-\beta\varepsilon_0(\mathbf{k})}) - t'\left((2\pi)^{-3}\int_{|k-k_F|>k_0}(e^{\beta\varepsilon_0(\mathbf{k})}+1)^{-1}\right)^2. \tag{11.17}$$

Here $\varepsilon_0(\mathbf{k}) = (k^2/2m) - \lambda$ is the nonperturbation energy spectrum, t' is the t-matrix defined below by (11.21). The renormalization $\lambda \to \mu$ is described by the formula.

$$\mu = \lambda - t'(2\pi)^{-3}\int_{|k-k_F|>k_0} d^3k(e^{\beta\varepsilon_0(\mathbf{k})}+1)^{-1}. \tag{11.18}$$

Finally, the t-matrix t' can be obtained as the result of summation of diagram sequence (10.13) which describes the scattering of opposite-spin particles, whereas the momenta corresponding to the internal lines of these diagrams acquire values outside the interval $|k-k_F|<k_0$. Such a t-matrix is given by the equation

$$t'(\mathbf{k}_1,\mathbf{k}_2;0) + (2\pi)^{-3}\int_{|k_3-k_F|>k_0} v(\mathbf{k}_1-\mathbf{k}_3)\frac{\tanh\beta\varepsilon_0(\mathbf{k}_3)/2}{2\varepsilon_0(\mathbf{k}_3)} t'(\mathbf{k}_3,\mathbf{k}_2;0)d^3k_3 = v(\mathbf{k}_1-\mathbf{k}_3). \tag{11.19}$$

This equation can be rewritten in the form

$$t'(\mathbf{k}_1,\mathbf{k}_2;0) + (2\pi)^{-3}\int v(\mathbf{k}_1-\mathbf{k}_3)t'(\mathbf{k}_3,\mathbf{k}_2;0)\frac{m}{k_3^2}d^3k_3 = v(\mathbf{k}_1-\mathbf{k}_3) + (2\pi)^{-3}\int v(\mathbf{k}_1-\mathbf{k}_3)t'(\mathbf{k}_3,\mathbf{k}_2;0) \times\left(\frac{m}{k_3^2} - \frac{\tanh\beta\varepsilon_0(\mathbf{k}_3)/2}{2\varepsilon_0(\mathbf{k}_3)}\right)d^3k_3, \tag{11.20}$$

and the operator acting on t' on the left-hand side can be inverted. This trick, which has been used many times above, allows us to determine the function $t'(\mathbf{k}_1,\mathbf{k}_2;0)$ which can be replaced by the constant

$$t' = t_0 + \frac{mt_0^2}{(2\pi)^3}\int_{|k-k_F|>k_0} d^3k\left(\frac{m}{k^2} - \frac{\tanh\beta\varepsilon_0(\mathbf{k})/2}{2\varepsilon_0(\mathbf{k})}\right) = t_0 + \frac{mt_0^2k_F}{2\pi^2}\left(2 - \ln\frac{2k_F}{k_0}\right) \tag{11.21}$$

in the domain $k \sim k_F$. This constant depends on the auxiliary parameter k_0. However, it will become evident from what follows that the final results are k_0-independent. Let us make two remarks concerning formula (11.16).

(1) The incorporation of equal-spin particles into the description of scattering is equivalent to the replacement of the antisymmetric potential by the antisymmetric t-matrix vanishing if t' is replaced by a constant. Thus, the interaction of equal-spin particles does not contribute to $\tilde{S}$ in the first approximation.

(2) We have neglected the $\mathbf{A}$ and φ dependences of p_0. The reasonability of this procedure will be discussed below.

The effect of superconductivity should demonstrate itself after the integration of the functional $\exp \tilde{S}$ over slow variables ψ_0 and $\bar{\psi}_0$. Let us proceed from the integration over Fermi fields to the integration over auxiliary Bose fields. This completes the analogy between superconductivity and superfluidity. To accomplish this idea we insert the Gaussian integral

$$\int \mathrm{D}\bar{c}\mathrm{D}c \exp\left(\int \mathrm{d}\tau \mathrm{d}^3x (t')^{-1} \bar{c}(\mathbf{x},\tau) c(\mathbf{x},\tau)\right) \tag{11.22}$$

over an auxiliary Bose field $c(\mathbf{x},\tau)$ into the integral over Fermi fields ψ_0 and $\bar{\psi}_0$. If the interaction potential $u(\mathbf{x}-\mathbf{y})$ is assumed to be attractive and the quantity t' negative, then the quadratic form in the integrand of (11.22) is negative. After the translation of integration variables

$$\begin{aligned} c(\mathbf{x},\tau) &\to c(\mathbf{x},\tau) + t'\psi_{0+}(\mathbf{x},\tau)\psi_{0-}(\mathbf{x},\tau), \\ \bar{c}(\mathbf{x},\tau) &\to \bar{c}(\mathbf{x},\tau) + t'\bar{\psi}_{0-}(\mathbf{x},\tau)\bar{\psi}_{0+}(\mathbf{x},\tau) \end{aligned} \tag{11.23}$$

the integrand of the integral over $c, \bar{c}$ transforms into $\exp \tilde{\tilde{S}}$, where

$$\begin{aligned} \tilde{\tilde{S}} = \int \mathrm{d}\tau \mathrm{d}^3x \Big\{ &\sum_s (\bar{\psi}_{0s}\partial_\tau \psi_{0s} - (2m)^{-1} |(\nabla + ie\mathbf{A})\psi_{0s}|^2 \\ &+ (\mu + ie\varphi + (seH/2m))\bar{\psi}_{0s}\psi_{0s} + \psi_{0+}\psi_{0-}\bar{c} \\ &+ \bar{\psi}_{0-}\bar{\psi}_{0+}c + (t')^{-1}|c|^2 \Big\}. \end{aligned} \tag{11.24}$$

It is quadratic in $\psi_{0s}, \bar{\psi}_{0s}$. Now it is possible to calculate the Gaussian integral over $\psi_{0s}, \bar{\psi}_{0s}$. It is formally equal to the determinant of the operator

$$M = \begin{pmatrix} \partial_\tau + \dfrac{eH}{2m} - \dfrac{1}{2m}(\nabla + ie\mathbf{A})^2 + \mu + ie\varphi, & -\bar{c} \\ -c, & \partial_\tau + \dfrac{eH}{2m} + \dfrac{1}{2m}(\nabla - ie\mathbf{A})^2 - \mu - ie\varphi \end{pmatrix} \tag{11.25}$$

acting on the pair of Fermi fields ψ_{0+}, ψ_{0-}. The operator depends on the functions $\varphi, \mathbf{A}, c$ and $\bar{c}$. Its determinant can be regularized by dividing it by the determinant of the operator M_0 corresponding to the zero values of the variables $\varphi, \mathbf{A}, c, \bar{c}$. Thus, we can write

$$\int \mathrm{D}\bar{\psi}_0 \mathrm{D}\psi_0 \exp \tilde{\bar{S}} = \exp\left(\int \mathrm{d}\tau\, \mathrm{d}x (t')^{-1} |c|^2 + \ln \det M/M_0\right). \quad (11.26)$$

Let us evaluate $\ln \det M/M_0$ according to the formula

$$\ln \det M/M_0 = \ln \det M/M_1 + \ln \det M_1/M_0, \quad (11.27)$$

where M_1 is the operator M (11.25) at $c = \bar{c} = 0$ (but A, φ do not vanish). The second term in (11.27) does not depend on the Bose field $c(\mathbf{x}, \tau)$, it will be considered later. The first term can be expanded into the functional series $c, \bar{c}$:

$$\ln \det M/M_1 = \mathrm{Tr} \ln M/M_1 = \mathrm{Tr} \ln (I - Gu)$$

$$= -\sum_{n=1}^{\infty} \frac{1}{n} \int \sum_{i=1}^{n} \mathrm{d}^3 x_i\, \mathrm{d}\tau_i\, \mathrm{tr}\,(G(\mathbf{x}_1, \tau_1; \mathbf{x}_2, \tau_2 | \varphi, A) u(\mathbf{x}_2, \tau_2)$$

$$\times G(\mathbf{x}_2, \tau_2; \mathbf{x}_3, \tau_3 | \varphi, \mathbf{A}) u(\mathbf{x}_3, \tau_3) \cdots u(\mathbf{x}_1, \tau_1)). \quad (11.28)$$

Here, the symbol tr denotes the range-two matrix trace; $u(\mathbf{x}, \tau)$ is the matrix

$$u(\mathbf{x}, \tau) = \begin{pmatrix} 0, & c(\mathbf{x}, \tau) \\ \bar{c}(\mathbf{x}, \tau), & 0 \end{pmatrix}, \quad (11.29)$$

and G is the Green function matrix

$$G = \begin{pmatrix} G_+ & 0 \\ 0 & G_- \end{pmatrix} \quad (11.30)$$

in which $G_\pm$ are the Green functions defined by the equations

$$\left(\partial_\tau + \frac{eH}{2m} \mp \left(\frac{1}{2m}(\nabla \pm ie\mathbf{A})^2 - \mu - ie\varphi\right)\right) G_\pm(\mathbf{x}, \tau; \mathbf{y}, \tau_1 | \varphi, \mathbf{A})$$

$$= \delta(\mathbf{x} - \mathbf{y}) \delta(\tau - \tau_1). \quad (11.31)$$

Only the terms with even n contribute to expansion (11.28), as G is a diagonal matrix and the diagonal elements of u are equal to zero.

We restrict ourselves to terms of the second and fourth order in $c(\mathbf{x}, \tau)$ in expansion (11.28). This can be done near the phase transition point, where, as will be shown below, the leading contribution to the integral over $|c(\mathbf{x}, \tau)|^2$ originates from a neighbourhood of a positive constant ρ_0 which is the condensate density that vanishes in the limit $T \to T_c$.

We neglect the dependence of Green's functions $G_{\pm}$ on the fields φ and A in the fourth-order term and assume that in the first approximation the arguments of all functions u are identical. This results in the following expression of the fourth-order term:

$$-\frac{1}{2}\int \prod_{i=1}^{4} d\tau_i d\mathbf{x}_i |c(\mathbf{x}_1, \tau_1)|^2 G_+(1,2)G_-(2,3)G_+(3,4)G_-(4,1), \quad (11.32)$$

where

$$G_{\pm}(1,2) = G_{\pm}(\mathbf{x}_1, \tau_1; \mathbf{x}_2, \tau_2).$$

Proceeding to the momentum-representation of Green's functions $G_{\pm}$, we are able to perform the integration over $(\mathbf{x}_2, \tau_2)$, $(\mathbf{x}_3, \tau_3)$, $(\mathbf{x}_4, \tau_4)$. This leads to the formula

$$-\frac{b}{2}\int d\tau d^3x |c(\mathbf{x}, \tau)|^4 \quad (11.33)$$

for (11.32). The coefficient

$$b = (\beta V)^{-1} \sum_{p} (\omega^2 + \xi^2(\mathbf{k}))^{-2} \quad (11.34)$$

can easily be calculated by integrating (in the interval $|k - k_F| < k_0$) over $\xi(\mathbf{k}) = (k^2/2m) - \mu$:

$$b = mk_F \frac{T}{2\pi^2} \sum_{\omega} \int \frac{d\xi}{(\xi^2 + \omega^2)^2}$$

$$= \frac{mk_F T}{2\pi^2} \frac{\pi}{2} \sum_{\omega} |\omega|^{-3} = \frac{7\zeta(3) mk_F}{16\pi^4 T^2}, \quad (11.35)$$

where $\zeta(3)$ is the Riemann ζ function for $s = 3$ and T in (11.35) should be put equal to the phase transition temperature T_c (10.31) (because this procedure is valid in the vicinity of T_c).

The second-order term in expansion (11.28) has the form

$$-\int d\mathbf{x} d\mathbf{y} d\tau d\tau_1 G_+(\mathbf{x}, \tau; \mathbf{y}, \tau_1 | \varphi, \mathbf{A}) G_-(\mathbf{y}, \tau_1; \mathbf{x}, \tau | \varphi, \mathbf{A})$$
$$\times \bar{c}(\mathbf{x}, \tau) c(\mathbf{y}, \tau_1). \quad (11.36)$$

In order to find the $\varphi, \mathbf{A}$-dependence of this term we use the following approximation:

$$G_{\pm}(\mathbf{x}, \tau; \mathbf{y}, \tau_1 | \varphi, \mathbf{A})$$
$$\approx \exp\left(\pm ie \int_{(\mathbf{x},\tau)}^{(\mathbf{y},\tau_1)} (\mathbf{A} d\mathbf{x} + \varphi d\tau) \right) G_{0\pm}(\mathbf{x} - \mathbf{y}, \tau - \tau_1) \quad (11.37)$$

for the Green function of a charged particle in a slow varying electromagnetic field. The integral on the right-hand side of (11.37) is taken along the straight line connecting $(\mathbf{x}, \tau)$ and $(\mathbf{y}, \tau_1)$.

Employing the relation

$$c(\mathbf{y}, \tau_1) = \exp\left(\mathrm{i} \int_{(\mathbf{x},\tau)}^{(\mathbf{y},\tau_1)} (\nabla \, \mathrm{d}\mathbf{x} + \partial_\tau \, \mathrm{d}\tau) \right) c(\mathbf{x}, \tau) \tag{11.38}$$

one can rewrite (11.36) in the form

$$-\int \mathrm{d}\mathbf{x} \, \mathrm{d}\mathbf{y} \, \mathrm{d}\tau \, \mathrm{d}\tau_1 \, G_{0+}(\mathbf{x}-\mathbf{y}, \tau - \tau_1) G_{0-}(\mathbf{y}-\mathbf{x}, \tau_1 - \tau) \bar{c}(\mathbf{x}, \tau)$$
$$\times \exp\left(\mathrm{i} \int_{(\mathbf{x},\tau)}^{(\mathbf{y},\tau_1)} (\nabla + 2\mathrm{i}e\mathbf{A}) \, \mathrm{d}\mathbf{x} + (\partial_\tau + 2\mathrm{i}e\varphi) \, \mathrm{d}\tau \right) c(\mathbf{x}, \tau). \tag{11.39}$$

From this formula it can be seen that in order to calculate the value of this functional, one has first to calculate the functional for vanishing external fields and then to replace the derivatives of $c(\mathbf{x}, \tau)$ by the 'covariant derivatives' according to the rule

$$\nabla \to \nabla + 2\mathrm{i}e\mathbf{A}, \quad \partial_\tau \to \partial_\tau + 2\mathrm{i}e\varphi. \tag{11.40}$$

Functional (11.36) for $\varphi = 0$ and $\mathbf{A} = 0$ can most easily be calculated in p representation, where it is equal to

$$-\sum_p A(p) c^*(p) c(p). \tag{11.41}$$

Here

$$A(p) = -(\beta V)^{-1} \sum_p (\mathrm{i}\omega_1 - \xi(\mathbf{k}_1))^{-1} (\mathrm{i}\omega - \mathrm{i}\omega_1 - \xi(\mathbf{k} - \mathbf{k}_1))^{-1}$$
$$= -(2\pi)^{-3} \int_{|k_1 - k_F| < k_0} \frac{\tanh \beta\xi(\mathbf{k}_1)/2 + \tanh \beta\xi(\mathbf{k} - \mathbf{k}_1)/2}{2[\xi(\mathbf{k}_1) + \xi(\mathbf{k} - \mathbf{k}_1) - \mathrm{i}\omega]} \mathrm{d}^3 k_1, \tag{11.42}$$

where

$$\xi(\mathbf{k}) = (k^2/2m) - \mu.$$

Let us extract the constant $A(0)$ from the $A(p)$ and add the contribution of this constant to $(t')^{-1} \int \mathrm{d}\tau \, \mathrm{d}\mathbf{x} |c|^2$. For $\varphi = 0$ and $\mathbf{A} = 0$ we obtain the expression

$$\sum_p (A(0) - A(p)) c^*(p) c(p) + ((t')^{-1} - A(0))$$
$$\times \int \mathrm{d}\tau \, \mathrm{d}\mathbf{x} |c(\mathbf{x}, \tau)|^2 - \frac{b}{2} \int \mathrm{d}\tau \, \mathrm{d}\mathbf{x} |c(\mathbf{x}, \tau)|^4. \tag{11.43}$$

The functional resembles the action functional of a nonideal Bose gas. The first component is an analogue of the kinetic term, and the third component describes the self-interaction of the Bose field $c(x, \tau)$. By using (11.21), the coefficient of the second term can be written in the form

$$t_0^{-1} - \frac{mk_F}{2\pi^2}\left(2 - \ln\frac{2k_F}{k_0}\right) + (2\pi)^{-3}\int_{|k-k_F|<k_0} d^3k\,\frac{\tanh\beta\xi(\mathbf{k})/2}{2\xi(\mathbf{k})}$$
$$= t_0^{-1} + \frac{mk_F}{2\pi^2}\left(-2 + \ln\frac{2k_F}{k_0} + \int_0^{k_Fk_0/m}\frac{d\xi}{\xi}\tanh\frac{\beta}{2}\xi\right). \tag{11.44}$$

This expression vanishes at $T = T_c$ by virtue of (10.28). It can therefore be rewritten in the form

$$\frac{mk_F}{2\pi^2}\int_0^\infty\frac{d\xi}{\xi}\left(\tanh\frac{\beta\xi}{2} - \tanh\frac{\beta_c\xi}{2}\right) = \frac{mk_F}{2\pi^2}\ln\frac{T_c}{T} \approx \frac{mk_F}{2\pi^2}\frac{\Delta T}{T_c}, \tag{11.45}$$

where $\Delta T = T_c - T$. For $T < T_c$, coefficient (11.44) is positive, and functional (11.43) has a maximum at $|c(\mathbf{x}, \tau)|^2 = \rho_0 = \text{const}$. This constant quantity

$$\rho_0 = \frac{mk_F}{2\pi^2}\frac{\Delta T}{T_c b} = \frac{8\pi^2 T_c \Delta T}{7\zeta(3)} \tag{11.46}$$

is the condensate density.

Thus, in the formalism developed so far, the phase transition of a Fermi system can be described in the same way as in the theory of Bose gas. The main contribution to the integral over the auxiliary Bose field $c(\mathbf{x}, \tau)$ is given by functions with the modulus squared $|c|^2$ close to the condensate density ρ_0.

In the following, functional (11.43) will be applied only to the description of stationary effects in superconductors. To proceed we integrate over the Fourier coefficients $c(p)$, $c^*(p)$, with $\omega \neq 0$. This yields a constant additional term to action (11.43) and leads to the renormalization of coefficients in the second- and fourth-order forms in (11.49). One can show that these corrections are small and may be neglected in the first approximation. Thus, up to an unessential constant term, the result of integration over $c(p)$, $c^*(p)$, $\omega \neq 0$, is that one can consider only the τ-independent functions $c(\mathbf{x}, \tau)$, $\bar{c}(\mathbf{x}, \tau)$ with Fourier coefficients having the frequency $\omega = 0$ in (11.43).

Finally, we obtain the functional

$$\beta\sum_{\mathbf{k}}(A(0) - A(\mathbf{k}))c^*(\mathbf{k})c(\mathbf{k}) + \frac{\beta mk_F\Delta T}{2\pi^2 T_c}\int d^3x\,|c(\mathbf{x})|^2 - \frac{b\beta}{2}\int d^3x\,|c(\mathbf{x})|^4 \tag{11.47}$$

instead of (11.43). For small k we have

$$A(0)-A(\mathbf{k})$$

$$=(2\pi)^{-3}\int d^3k_1 \frac{\tanh\frac{\beta}{2}\xi\left(\mathbf{k}_1+\frac{\mathbf{k}}{2}\right)+\tanh\frac{\beta}{2}\xi\left(\mathbf{k}_1-\frac{\mathbf{k}}{2}\right)}{2\left(\xi\left(\mathbf{k}_1+\frac{\mathbf{k}}{2}\right)+\xi\left(\mathbf{k}_1-\frac{\mathbf{k}}{2}\right)\right)}-\frac{\tanh\frac{\beta}{2}\xi(\mathbf{k}_1)}{2\xi(\mathbf{k}_1)}$$

$$\approx(2\pi)^{-3}\int d^3k_1 \frac{\left(\tanh\frac{\beta}{2}\xi\right)''}{4\xi}\left(\frac{\beta(\mathbf{k}_1\mathbf{k})}{4m}\right)^2$$

$$=\frac{k^2k_F^3}{384\pi^2mT_c^2}\int\frac{dx}{x}(\tanh x)'', \tag{11.48}$$

where

$$\int\frac{dx}{x}(\tanh x)''=-\frac{32}{\pi^2}\sum_{n=0}^{\infty}(2n+1)^{-3}=-\frac{28}{\pi^2}\zeta(3). \tag{11.49}$$

Finally, the kinetic term in (11.47) takes the form

$$\beta\sum_{\mathbf{k}}(A(0)-A(\mathbf{k}))|c(\mathbf{k})|^2=-\frac{\beta k_F\varepsilon_F 7\zeta(3)}{48\pi^4T_c^2}\int d^3x|\nabla c(\mathbf{x})|^2, \tag{11.50}$$

where $\varepsilon_F=\mu$ is the Fermi energy.

We now introduce the new function

$$\psi(\mathbf{x})=\left(\frac{7\zeta(3)mk_F\varepsilon_F}{12\pi^4T_c^2}\right)^{1/2}c(\mathbf{x}) \tag{11.51}$$

instead of $c(\mathbf{x})$ and write down functional (11.47) replacing the derivative ∇ in the kinetic term by the covariant derivative $\nabla+2ie\mathbf{A}$ according to (11.40):

$$\beta\int d^3x\left(-\frac{1}{4m}|(\nabla+2ie\mathbf{A})\psi(\mathbf{x})|^2+\Lambda|\psi(\mathbf{x})|^2-\frac{g}{2}|\psi(\mathbf{x})|^4\right). \tag{11.52}$$

Here

$$\Lambda=\frac{6\pi^2T_c\Delta T}{7\zeta(3)\varepsilon_F},\quad g=\frac{9\pi^4T_c^2}{7\zeta(3)mk_F\varepsilon_F^2}. \tag{11.53}$$

The expression

$$-\frac{\beta}{2}\int d^3x(\operatorname{curl}\mathbf{A})^2 \tag{11.54}$$

describing the stationary magnetic field, as well as the contribution of $\ln\det M_1/M_0$ from (11.27) should be added to functional (11.52). This functional describes the magnetic polarization of a medium. For weak fields $\mathbf{A}(\mathbf{x})$ it differs from (11.54) by the polarization coefficient κ. The coefficient is of order $\alpha v_F/c$, where $\alpha = 1/137$ is the fine structure constant, v_F/c is the ratio of the electron velocity on the Fermi surface to the velocity of light. As $\kappa \ll 1$, one may neglect the contribution of $\ln\det M_1/M_0$ in the first approximation and consider the functional

$$S_{\text{eff}} = -\beta\int d^3x\left(\frac{1}{2}(\operatorname{curl}\mathbf{A})^2 + \frac{1}{4m}|(\nabla + 2ie\mathbf{A})\psi(\mathbf{x})|^2 - \Lambda|\psi(\mathbf{x})|^2 + \frac{g}{2}|\psi(\mathbf{x})|^4\right). \tag{11.55}$$

The functional $\exp S_{\text{eff}}$ should be integrated over the complex functions $\psi(\mathbf{x}), \bar{\psi}(\mathbf{x})$ as well as over the vector potential $\mathbf{A}(\mathbf{x})$. The leading contribution to this integral is given by values close to the classical solutions. These solutions satisfy the system of equations

$$-\frac{1}{4m}(\nabla + 2ie\mathbf{A})^2\psi(\mathbf{x}) - \Lambda\psi(\mathbf{x}) + g|\psi(\mathbf{x})|^2\psi(\mathbf{x}) = 0,$$

$$\operatorname{curl}\operatorname{curl}\mathbf{A} = j = -\frac{ie}{2m}(\nabla\bar{\psi}\psi - \bar{\psi}\nabla\psi) - \frac{2e^2}{m}\bar{\psi}\psi\mathbf{A} \tag{11.56}$$

which are identical to the equations of Ginsburg and Landau.

The functional integration method employed in the derivation of these equations also allows us, in principle, to describe the fluctuations of the field $\psi, \bar{\psi}, \mathbf{A}$ around the classical solutions. To account for these fluctuations, one has to shift the solutions of classical equations

$$\psi \to \psi + \psi_0, \quad \bar{\psi} \to \bar{\psi} + \bar{\psi}_0, \quad \mathbf{A} \to \mathbf{A} + \mathbf{A}_0 \tag{11.57}$$

in the functional integral and to construct a perturbation theory where the quadratic form of the fields $\psi, \bar{\psi}$ and $\mathbf{A}$ represents the nonperturbed action and higher-order forms are regarded as a perturbation. This will not be accomplished here.

As an example of the application of Ginsburg–Landau equations, we shall determine the vortex lattice structure near the phase transition and the value of the critical magnetic field $\mathbf{H}$ (usually denoted as H_{c_2}). The quantity H_{c_2} is the field which must be applied to a superconductor so as to cause its transition to the normal state. The right-hand side of the second equation in

(11.56) and the term $g|\psi|^2\psi$ in the first equation may be neglected near the phase transition. The second equation has the solution

$$\mathbf{A} = \tfrac{1}{2}[\mathbf{H}, \mathbf{x}].$$

The first equation

$$-\frac{1}{4m}(\nabla + ie[\mathbf{H}, \mathbf{x}])^2\psi(\mathbf{x}) = \Lambda\psi(\mathbf{x}) \tag{11.58}$$

is the Schrödinger equation for a particle with charge $2e$ and mass $2m$ in a constant magnetic field H. If the solution does not depend on the component of $\mathbf{x}$ parallel to $\mathbf{H}$, the spectrum of possible values of Λ (i.e. the Landau levels) has the form

$$\Lambda = \frac{eH}{m}(n + \tfrac{1}{2}), \quad n = 0, 1, 2, \ldots \tag{11.59}$$

The smallest value of Λ ($n = 0$) corresponds to the highest value of the critical field

$$H = H_{c_2} = \frac{2m\Lambda}{e} = \frac{12\pi^2 m T_c \Delta T}{7\zeta(3)e\varepsilon_F}. \tag{11.60}$$

Let us construct the solution of (11.58) describing a periodic structure. We take a particular solution $\varphi(\mathbf{x})$ corresponding to the lowest Landau level. The function $\varphi(\mathbf{x} + \mathbf{a})$ satisfies equation (11.58) with $[\mathbf{H}, \mathbf{x}]$ replaced by $[\mathbf{H}, \mathbf{x}] + [\mathbf{H}, \mathbf{a}]$, which corresponds to the gauge transformation of the vector potential. One can return to the previous gauge by transforming $\varphi \to \varphi \exp(-ie([\mathbf{H}, \mathbf{a}], \mathbf{x}))$.

Thus, the function $\varphi(\mathbf{x} + \mathbf{a}) \exp(-ie([\mathbf{H}, \mathbf{a}], \mathbf{x}))$ satisfies (11.58), too. Putting $\mathbf{a}$ equal to the lattice vector and performing the summation over all such vectors, we obtain the function

$$\psi(\mathbf{x}) = \sum_{\mathbf{a}} \varphi(\mathbf{x} + \mathbf{a}) \exp(-ie([\mathbf{H}, \mathbf{a}], \mathbf{x})) \tag{11.61}$$

describing a periodic structure. We recall here that two arbitrary lattice vectors have to satisfy condition (11.11). Using this equation, we obtain the transformation law for $\psi(\mathbf{x})$ shifted by the lattice vector $\mathbf{b}$:

$$\psi(\mathbf{x} + \mathbf{b}) = \psi(\mathbf{x}) \exp(ie([\mathbf{H}, \mathbf{b}], \mathbf{x})). \tag{11.62}$$

It is suitable to take the function

$$\varphi(\mathbf{x}) = (x - iy) \exp\left(-\frac{eH}{2}(x^2 + y^2)\right), \tag{11.63}$$

where x, y are the coordinates in the plane perpendicular to H, as the particular solution corresponding to the lowest Landau level. The function (11.61) corresponding to this solution acquires zero values in the lattice sites, and its phase acquires an increment 2π per passage around these points. One can therefore say that (11.61) describes the quantum vortex lattice formed in a superconductor under the influence of a magnetic field less than the critical value H_{c_2} (11.60). One can show (considering also the nonlinear terms in (11.56)) that the energy is minimal for a triangular lattice.

The periodic quantum vortex lattice also arises in a rotating superfluid Bose system, where the role of the magnetic field $\mathbf{H}$ is assumed by the angular velocity vector $\boldsymbol{\omega}$. The method of description of a vortex lattice developed in this section is also applicable to Bose systems.

12
Collective excitations in superfluid Fermi systems

The perturbative scheme outlined in section 10 for a system of Fermi particles in the superconducting state is not quite sufficient for the description of collective Bose excitations in such a system. We are going to demonstrate that the idea of integration over fast and slow fields, together with the introduction of new Bose field variables instead of Fermi ones, provides a method for describing collective excitations in Fermi superfluids. The phase transition of a Fermi system into the superfluid state will be described as the Bose condensation of some Bose system in this formalism. The method may be applied to the case of a pairing in the s-state superconductor model as well as to the case of the p-pairing (^{3}He model).

The starting point is the action functional for a system of Fermi particles with spin

$$S = \int_0^\beta d\tau \int dx \sum_s \bar{\psi}_s(\mathbf{x},\tau)\partial_\tau \psi_s(\mathbf{x},\tau) - \int_0^\beta H'(\tau)\,d\tau, \tag{12.1}$$

where

$$\begin{aligned} H'(\tau) = \int dx \sum_s \bigg(& \frac{1}{2m} \nabla\bar{\psi}_s(\mathbf{x},\tau)\nabla\psi_s(\mathbf{x},\tau) \bigg) \\ & - (\lambda + s\mu_0 H)\bar{\psi}_s(\mathbf{x},\tau)\psi_s(\mathbf{x},\tau) + \frac{1}{2}\int dx\,dy\,u(\mathbf{x}-\mathbf{y}) \\ & \times \sum_{s,s'} \bar{\psi}_s(\mathbf{x},\tau)\bar{\psi}_{s'}(\mathbf{y},\tau)\psi_{s'}(\mathbf{y},\tau)\psi_s(\mathbf{x},\tau). \end{aligned} \tag{12.2}$$

Here

$$\psi_s(\mathbf{x},\tau) = (\beta V)^{-1/2} \sum_{\mathbf{k},\omega} e^{i(\omega\tau + \mathbf{kx})} a_s(p) \tag{12.3}$$

$\omega = (2n+1)\pi/\beta$, $s = \pm$ is the spin index, μ_0 is the (bare) magnetic moment of a particle, H is a magnetic field.

First of all we integrate over the fast Fermi fields ψ_{1s}, $\bar{\psi}_{1s}$, such that their Fourier components $a_s(p)$, $a_s^*(p)$ have $|k - k_F| > k_0$ or $|\omega| > \omega_0$. The parameters k_0 and ω_0 are only defined up to their order of magnitude. All the physical results must not depend on the specific choice of k_0 and ω_0. So we have

$$\int D\bar{\psi}_{1s} D\psi_{1s} \exp S = \exp \tilde{S}[\psi_{0s}, \bar{\psi}_{0s}], \tag{12.4}$$

where $\tilde{S}[\psi_{0s}, \bar{\psi}_{0s}]$ is the effective action for the slow fields $\psi_{0s}, \bar{\psi}_{0s}$ which only have the Fourier components for which $|k - k_F| < k_0$ and $|\omega| < \omega_0$.

The most general form of $\tilde{S}$ is

$$\tilde{S} = \sum_{n=0}^{\infty} \tilde{S}_{2n}, \tag{12.5}$$

where $\tilde{S}_{2n}$ is the form of order $2n$ in the fields $\psi_{0s}, \bar{\psi}_{0s}$.

The constant term $\tilde{S}_0$ in (12.5) will be irrelevant in the further considerations. The higher-order terms $\tilde{S}_6, \tilde{S}_8$ and so on in (12.5) may be neglected when the 'low-energy shell' $|k - k_F| < k_0$ is thin. The most general form of $\tilde{S}_2$ is

$$\tilde{S}_2 = \sum_{\substack{s, |\omega| < \omega_0 \\ |k - k_F| < k_0}} \varepsilon_s(\mathbf{k}, \omega, H) a_s^*(p) a_s(p). \tag{12.6}$$

Supposing that $\varepsilon_s(k = k_F, \omega = 0, H = 0) = 0$ we can use the following form of ε_s:

$$\varepsilon_s(\mathbf{k}, \omega, H) = Z^{-1}(i\omega - c_F(k - k_F) + s\mu H). \tag{12.7}$$

We expand ε_s into a series in $\omega, k - k_F, H$ and take into account only the linear terms. Here c_F is the velocity on the Fermi sphere, μ is the magnetic moment of a Fermi quasiparticle and Z is the normalization constant.

The form $\tilde{S}_4$ describes the pair interaction of quasiparticles and is equal to

$$-\frac{1}{\beta V} \sum_{p_1 + p_2 = p_3 + p_4} t_0(p_1, p_2, p_3, p_4) a_+^*(p_1) a_-^*(p_2) a_-(p_4) a_+(p_3)$$

$$-\frac{1}{2\beta V} \sum_{p_1 + p_2 = p_3 + p_4} t_1(p_1, p_2, p_3, p_4)(2a_+^*(p_1) a_-^*(p_2) a_-(p_4) a_+(p_3)$$

$$+ a_+^*(p_1) a_+^*(p_2) a_+(p_4) a_+(p_3) + a_-^*(p_1) a_-^*(p_2) a_-(p_4) a_-(p_3)). \tag{12.8}$$

Here $t_0(p_i)$ is a scattering amplitude symmetric under the permutations $p_1 \rightleftarrows p_2$ or $p_3 \rightleftarrows p_4$; $t_1(p_i)$ is antisymmetric under the same permutations.

Near the Fermi sphere we may put $\omega_i = 0, \mathbf{k}_i = k_F \mathbf{n}_i$ in $t_0(p_i), t_1(p_i)$, where $\mathbf{n}_i$ are unit vectors obeying the following condition

$$\mathbf{n}_1 + \mathbf{n}_2 = \mathbf{n}_3 + \mathbf{n}_4.$$

The scattering amplitudes t_0, t_1 must depend only on two invariants, for instance on $(\mathbf{n}_1, \mathbf{n}_2)$ and $(\mathbf{n}_1 - \mathbf{n}_2, \mathbf{n}_3 - \mathbf{n}_4)$. So we have

$$t_0 = f((\mathbf{n}_1, \mathbf{n}_2), (\mathbf{n}_1 - \mathbf{n}_2, \mathbf{n}_3 - \mathbf{n}_4)),$$
$$t_1 = (\mathbf{n}_1 - \mathbf{n}_2, \mathbf{n}_3 - \mathbf{n}_4) g((\mathbf{n}_1, \mathbf{n}_2), (\mathbf{n}_1 - \mathbf{n}_2, \mathbf{n}_3 - \mathbf{n}_4)). \qquad (12.9)$$

Both f and g are even functions of the second variable.

The functional $\tilde{S}_2 + \tilde{S}_4$ is the most general expression which describes Fermi quasiparticles near the Fermi surface and their pair interaction.

The functions f and g can be calculated for low-density systems. For realistic systems they must be defined experimentally. The approach outlined for obtaining $\tilde{S}_2 + \tilde{S}_4$ in the functional integral formalism is an alternative to the Landau theory of the Fermi liquid (Landau, 1956, 1957).

We will consider below the two simplest cases

$$f = f_0 = \text{const.} < 0, \quad g = 0 \text{ 'superconductor'}$$
$$f = 0, \quad g = g_0 = \text{const.} < 0 \text{ '}^3\text{He model'}. \qquad (12.10)$$

In both cases we can go from the integral over Fermi fields to the integral over some auxiliary Bose field. In order to do so, we insert a Gaussian integral of $\exp(\bar{c}Ac)$ (A is some operator) over the fields $c, \bar{c}$ into the integral over the Fermi field and than perform the shift transformation which cancels the $\tilde{S}_4$ functional. After this transformation we obtain a Gaussian integral over the Fermi fields which can be written in closed form. We obtain the following effective action functional:

$$S_{\text{eff}}[c, \bar{c}] = \bar{c}Ac + \ln \det M[c, \bar{c}]/M[0, 0]. \qquad (12.11)$$

This functional contains all the information of collective Bose excitations in a given Fermi system.

The Bose system described by functional (12.11) may undergo a phase transition into the superfluid state due to the Bose condensation. We can find the phase transition temperature T_c for both cases (12.10). Then we shall look for the Bose spectrum of quasiparticles for $T > T_c$, as well as for $T < T_c$.

First of all let us consider the simplest case $t_0 = f_0 < 0$, $t_1 = 0$. We introduce the following Gaussian integral:

$$\int \exp\left(f_0^{-1} \sum_p |c(p)|^2 \right) \prod_p dc^*(p)\, dc(p) \qquad (12.12)$$

into the integral over Fermi fields. This is an integral over the complex Bose field $c(x, \tau)$ with Fourier coefficients $c(p)$.

If we now make the shift transformation

$$c(p) \to c(p) + \frac{f_0}{(\beta V)^{1/2}} \sum_{p_1 + p_2 = p} a_+(p_1) a_-(p_2), \tag{12.13}$$

which reduces out the interaction term $\tilde{S}_4$ in (12.5), we obtain the following action which depends on both Fermi and Bose fields:

$$f_0^{-1} \sum_p c^*(p)c(p) + \sum_{p,s} Z^{-1}(i\omega - c_F(k - k_F)) a_s^*(p) a_s(p)$$
$$+ (\beta V)^{-1/2} \sum_{p_1 + p_2 = p_3} (c^*(p_3) a_+(p_1) a_-(p_2) + c(p_3) a_-^*(p_2) a_+^*(p_1)). \tag{12.14}$$

The quadratic form of the Fermi variables may be written down as

$$\sum_{p_1, p_2, a, b} \chi_a^*(p_1) M_{ab}(p_1, p_2) \chi_b(p_2), \tag{12.15}$$

where

$$\chi_1(p) = a_+^*(p), \quad \chi_2(p) = a_-(p) \tag{12.16}$$

can be regarded as entries of the column on which the M operator of the form

$$M(p_1, p_2) = \begin{pmatrix} Z^{-1}(i\omega - \xi)\delta_{p_1, p_2}, (\beta V)^{-1/2} c(p_1 + p_2) \\ -(\beta V)^{-1/2} c^*(p_1 + p_2), Z^{-1}(-i\omega + \xi)\delta_{p_1, p_2} \end{pmatrix} \tag{12.17}$$

acts. Here $\xi = c_F(k - k_F)$. It is now possible to evaluate the integral over the Fermi fields in closed form. We obtain the following effective action functional:

$$S_{eff} = f_0^{-1} \sum_p c^*(p)c(p) + \ln \det M(c, c^*)/M(0, 0), \tag{12.18}$$

where M is defined by (12.17).

Let us now focus on the Fermi system in which the short-range attractive interaction between Fermi particles is supplemented by the long-range Coulomb interaction. This is a more realistic model of superconductor than the one considered above. In this case the starting action functional differs from (12.1) by the addend

$$-\frac{e^2}{2} \int_0^\beta d\tau \int d^3x \, d^3y |\mathbf{x} - \mathbf{y}|^{-1} \sum_{s,s'} \bar{\psi}_s(\mathbf{x}, \tau) \bar{\psi}_{s'}(\mathbf{y}, \tau) \psi_{s'}(\mathbf{y}, \tau) \psi_s(\mathbf{x}, \tau). \tag{12.19}$$

We may introduce the integral over a new variable $\varphi(\mathbf{x}, \tau)$ (the field of electric potential) as we did in section 9:

$$\int D\varphi \exp\left(-(8\pi)^{-1} \int d\tau \, d^3x (\nabla \varphi(\mathbf{x}, \tau))^2 \right). \tag{12.20}$$

The Coulomb interaction term (12.19) can be eliminated by making a shift transformation

$$\varphi(\mathbf{x},\tau)\to\varphi(\mathbf{x},\tau)+ie\int d^3y|\mathbf{x}-\mathbf{y}|^{-1}\sum_s\bar{\psi}_s(\mathbf{y},\tau)\psi_s(\mathbf{y},\tau). \tag{12.21}$$

Introducing the Gaussian integral (12.12), and eliminating the short-range interaction term by means of the shift transformation (12.13), we obtain the action depending on the Fermi fields $a_s(p)$, $a_s^*(p)$ and the Bose fields $c(p)$, $c^*(p)$, $\varphi(p)$ (Fourier coefficients of the field $\varphi(\mathbf{x},\tau)$):

$$\begin{aligned}&f_0^{-1}\sum_p c^*(p)c(p)-(8\pi)^{-1}\sum_p k^2\varphi(p)\varphi(-p)\\&+\sum_{p,s}Z^{-1}(i\omega-c_F(k-k_F))a_s^*(p)a_s(p)\\&+(\beta V)^{-1/2}\sum_{p_1+p_2=p_3}(c^*(p_3)a_+(p_1)a_-(p_2)+c(p_3)a_-^*(p_2)a_+^*(p_1))\\&-ie(\beta V)^{-1/2}\sum_{s,p_1-p_2=p_3}\varphi(p_3)a_s^*(p_1)a_s(p_2).\end{aligned} \tag{12.22}$$

After the integration over Fermi fields we obtain the following effective action:

$$\begin{aligned}S_{\text{eff}}=f_0^{-1}\sum_p c^*(p)c(p)-(8\pi)^{-1}\sum_p k^2\varphi(p)\varphi(-p)\\+\ln\det M[c,c^*,\varphi]/M[0,0,0],\end{aligned} \tag{12.23}$$

where M is equal to

$$M=\begin{pmatrix}Z^{-1}(i\omega-\xi(\mathbf{k}))\delta_{p_1,p_2}-ie(\beta V)^{-1/2}\varphi(p_1-p_2), & (\beta V)^{-1/2}c(p_1+p_2)\\-(\beta V)^{-1/2}c^*(p_1+p_2), & Z^{-1}(-i\omega+\xi(\mathbf{k}))\delta_{p_1,p_2}+ie(\beta V)^{-1/2}\varphi(p_2-p_1)\end{pmatrix}. \tag{12.24}$$

Let us now proceed to the case $f=0$, $g_0<0$ (^{3}He model). Here we can reduce out the form S_4 by performing the shift transformation

$$\begin{aligned}c_{i1}(p)&\to c_{i1}(p)+\frac{g_0}{2(\beta V)^{1/2}}\sum_{p_1+p_2=p}(n_{1i}-n_{2i})(a_+(p_2)a_+(p_1)-a_-(p_2)a_-(p_1))\\c_{i2}(p)&\to c_{i2}(p)+\frac{ig_0}{2(\beta V)^{1/2}}\sum_{p_1+p_2=p}(n_{1i}-n_{2i})(a_+(p_2)a_+(p_1)+a_-(p_2)a_-(p_1))\\c_{i3}(p)&\to c_{i3}(p)+\frac{g_0}{(\beta V)^{1/2}}\sum_{p_1+p_2=p}(n_{1i}-n_{2i})a_-(p_2)a_+(p_1).\end{aligned} \tag{12.25}$$

The quadratic form of Fermi variables can be written as

$$K = \frac{1}{2} \sum_{p_1, p_2, a, b} \psi_a^*(p_1) M_{ab}(p_1, p_2) \psi_b(p_2), \tag{12.26}$$

where the variables $\psi_a(p)$ are defined by

$$\begin{aligned} \psi_1(p) &= a_+(p), \quad \psi_2(p) = -a_-(p), \\ \psi_3(p) &= a_-^*(p), \quad \psi_4(p) = a_+^*(p) \end{aligned} \tag{12.27}$$

and $M(p_1, p_2)$ is a 4×4 matrix:

$$M = \begin{pmatrix} Z^{-1}(\mathrm{i}\omega - \xi(\mathbf{k}) + \mu H \sigma_3)\delta_{p_1, p_2}, (\beta V)^{-1/2}(n_{1i} - n_{2i})\sigma_a c_{ia}(p_1 + p_2) \\ -(\beta V)^{-1/2}(n_{1i} - n_{2i})\sigma_a c_{ia}^*(p_1 + p_2), Z^{-1}(-\mathrm{i}\omega + \xi(\mathbf{k}) + \mu H \sigma_3)\delta_{p_1, p_2} \end{pmatrix} \tag{12.28}$$

Here σ_a, $a = 1, 2, 3$, are the Pauli matrices. Now we can integrate over Fermi fields using the formula

$$\int \mathrm{e}^K \mathrm{D}\bar{\psi}_{0s} \mathrm{D}\psi_{0s} = (\det M)^{1/2}. \tag{12.29}$$

So we derive the effective action functional of the form

$$S_{\mathrm{eff}} = g_0^{-1} \sum_{p, i, a} c_{ia}^*(p) c_{ia}(p) + \tfrac{1}{2} \ln \det M(c_{ia}, c_{ia}^*)/M(0, 0). \tag{12.30}$$

In the sections that follow we shall show how one can obtain information on the physical properties of the systems described by effective action functionals (12.18), (12.23) and (12.30).

13

Bose spectrum of superfluid Fermi gas

We begin with the functional

$$S_{\text{eff}} = f_0^{-1} \sum_p c^*(p)c(p) + \ln\det M/M_0 \tag{13.1}$$

describing collective excitations in Fermi gas with short-range attractive interaction. This functional was investigated by Andrianov & Popov (1976). The problem of Bose excitations in the BCS (Bardeen, Cooper, Schriffer) model was discussed by Bogoliubov, Tolmachev & Shirkov (1958). The functional integral approach to Bose excitations in Fermi gas near the phase transition temperature was considered by Svidzinski (1971).

$|\Delta T| \ll T_c$

First of all we shall find the phase transition point. Near the phase transition (in the Ginsburg–Landau region $|\Delta T| \ll T_c$) we may expand ln det as a power series in $c, \bar{c}$. Denoting

$$M(0,0) = G^{-1}, \quad M(c,\bar{c}) = G^{-1} + u, \tag{13.2}$$

we have

$$\ln\det M(c,\bar{c})M^{-1}(0,0) = \ln\det(I + Gu)$$

$$= \operatorname{Tr}\ln(I + Gu) = -\sum_{n=1}^{\infty} \frac{1}{2n} \operatorname{Tr}(Gu)^{2n}. \tag{13.3}$$

Only even powers of the ln expansion contribute to Tr ln.

Near the phase transition point we will only take into account the first two nonvanishing terms in (13.3), so that

$$S_{\text{eff}} = -\sum_p A_0(p)c^*(p)c(p) - (2\beta V)^{-1}$$

$$\times \sum_{p_1+p_2=p_3+p_4} B(p_i)c^*(p_1)c^*(p_2)c(p_3)c(p_4). \tag{13.4}$$

Here

$$A_0(p) = -f_0^{-1} - (\beta V)^{-1} \sum_{p_1} Z^2 (i\omega_1 - \xi_1)^{-1} (i\omega + i\omega_1 - \xi_2)^{-1}$$

$$= \frac{Z^2}{(2\pi)^3} \int_{|k_1 - k_F| < k_0} d^3 k_i \frac{\tanh\frac{\beta}{2}\xi_1 + \tanh\frac{\beta}{2}\xi_2}{2(i\omega - \xi_1 - \xi_2)} - f^{-1}. \quad (13.5)$$

In the second term in (13.4) we may substitute $B(0)$ for $B(p_i)$ and write down this term as

$$-\frac{g}{2}\int d\tau\, dx\, |c(\mathbf{x}, \tau)|^4, \quad (13.6)$$

where

$$g = B(0) = \frac{Z^4}{\beta V} \sum_p (\omega^2 + \xi^2)^{-2} = \frac{7\zeta(3) k_F^2 Z^4}{16\pi^4 c_F T^2}. \quad (13.7)$$

Now we can find the phase transition temperature from the equation

$$A_0(0) = 0 \quad (13.8)$$

or

$$\frac{Z^2}{(2\pi)^3} \int d^3 k \frac{\tanh\frac{\beta}{2}\xi}{2\xi} + f_0^{-1} = 0. \quad (13.9)$$

Substituting $d^3 k$ by $4\pi k^2\, dk \approx 4\pi k_F^2 c_F^{-1}\, d\xi$, we can rewrite (13.9) as

$$\frac{Z^2 k_F^2}{2\pi^2 c_F} \int_0^{c_F k_0} d\xi \frac{\tanh\frac{\beta\xi}{2}}{\xi} + f_0^{-1} = 0. \quad (13.10)$$

The integral in (13.10) depends on k_0 logarithmically. In order that $T_c = \beta_c^{-1}$ be independent of k_0, we have to suppose that f_0^{-1} depends on k_0 in the following way:

$$f_0^{-1} = f^{-1} - \frac{Z^2 k_F^2}{2\pi^2 c_F} \ln\frac{k_0}{k_F}, \quad (13.11)$$

where f does not depend on k_0. Substituting (13.11) into (13.10) and using the formula

$$\int_0^{c_F k_0} \frac{d\xi}{\xi} \tanh\frac{\beta\xi}{2} = \int_0^{c_F k_0/2T} dx \frac{\tanh x}{x} \approx \ln\frac{2 c_F k_0 \gamma}{\pi T}, \quad (13.12)$$

where $\ln\gamma = C$ is the Euler constant, we obtain

$$T_c = \frac{2\gamma}{\pi} c_F k_F \exp\left(-\frac{2\pi^2 c_F}{|f| Z^2 k_F^2}\right). \tag{13.13}$$

Now we are able to obtain the Bose spectrum near T_c.

(*a*) If $T > T_c$, the equation for the spectrum is

$$A_0(p) = A_0(0) + (A_0(p) - A_0(0)) = 0, \tag{13.14}$$

where the analytic continuation $i\omega \to E$ is to be made. Then

$$\begin{aligned} A_0(0) &\approx \frac{Z^2 k_F^2}{2\pi^2 c_F} \int_0^\infty \frac{d\xi}{\xi}\left(\tanh\frac{\beta_c \xi}{2} - \tanh\frac{\beta\xi}{2}\right) \\ &= \frac{Z^2 k_F^2}{2\pi^2 c_F} \ln\frac{\beta_c}{\beta} \approx \frac{Z^2 k_F^2}{2\pi^2 c_F}\frac{T - T_c}{T_c}, \end{aligned} \tag{13.15}$$

$$\begin{aligned} A_0(p) - A_0(0) &= \frac{Z^2}{2(2\pi)^3}\int d^3k_1 \left[\frac{\tanh\frac{\beta}{2}\xi_1}{\xi_1} + \frac{\tanh\frac{\beta}{2}\xi_1 + \tanh\frac{\beta}{2}\xi_2}{i\omega - \xi_1 - \xi_2}\right] \\ &\approx -\frac{Z^2 k_F^2 |\omega|}{16\pi c_F T} - \frac{7\zeta(3) k_F^2 c_F k^2 Z^2}{96\pi^4 m T^2}. \end{aligned} \tag{13.16}$$

Substituting (13.15) and (13.16) into (13.14) and changing $|\omega|$ to $-iE$ we obtain the following result for the Bose spectrum for $T > T_c$:

$$E(\mathbf{k}) = -i\frac{7\zeta(3) c_F^2 k^2}{6\pi^3 T_c} - i\frac{8}{\pi}(T - T_c). \tag{13.17}$$

This spectrum is purely imaginary. The value $E(0)$ goes to zero as $T \to T_c$ (the system becomes a superconductor).

(*b*) $T < T_c$. If $T < T_c$, there is a Bose condensate density. We can find it by substituting $c(p), c^*(p) \to (\beta V)^{1/2} c\delta_{p,0}$ into (13.4):

$$-\beta V\left[-A_0(0)|c|^2 + \frac{g}{2}|c|^4\right]. \tag{13.18}$$

Looking for a maximum of (13.18) we obtain the condensate density

$$\rho_0 = |c|^2 = -\frac{A_0(0)}{g} = \frac{k_F^2 Z^2}{2\pi^2 g c_F}\frac{T_c - T}{T_c} = \frac{8\pi^2 T_c (T_c - T)}{7\zeta(3) Z^2}. \tag{13.19}$$

Now let us perform the shift transformation

$$c(p) = b(p) + (\rho_0 \beta V)^{1/2}\delta_{p,0}, \quad c^*(p) = b^*(p) + (\rho_0 \beta V)^{1/2}\delta_{p,0} \tag{13.20}$$

and consider the quadratic form of the new variables

$$-\frac{1}{2}\sum_{p}(2A(p)b^*(p)b(p) + \bar{B}(p)b(p)b(-p) + B(p)b^*(p)b^*(-p)), \qquad (13.21)$$

where

$$A(p) = A_0(p) + 2g\rho_0, \quad B(p) = g\rho_0. \qquad (13.22)$$

The shift transformation (13.20) will be sufficient if we confine ourselves to considering only the quadratic form (13.21). In order to construct the perturbation theory (without singularities) it would be more suitable to go to the density-phase variables (see section 7).

The Bose spectrum corresponding to the form (13.20) can be found from the equation

$$\det\begin{pmatrix} A(p), B(p) \\ B(p), A(-p) \end{pmatrix} = A(p)A(-p) - B^2(p) = 0. \qquad (13.23)$$

According to (13.16) and (13.22) both $A(p)$ and $B(p)$ are even functions of p, and (13.23) splits into two equations

$$A(p) - B(p) = 0, \quad A(p) + B(p) = 0, \qquad (13.24)$$

which give two branches of the spectrum:

$$E_1(\mathbf{k}) = -\mathrm{i}\frac{7\zeta(3)c_F^2k^2}{6\pi^2 T_c},$$

$$E_2(\mathbf{k}) = -\mathrm{i}\frac{7\zeta(3)c_F^2k^2}{6\pi^2 T_c} - \mathrm{i}\frac{16}{\pi}(T_c - T). \qquad (13.25)$$

The first branch is defined by the equation $A - B = 0$. It begins from zero ($E_1(\mathbf{k}) \to 0$ if $\mathbf{k} \to 0$). The second one is defined by $A + B = 0$. It differs from the branch (13.17) existing for $T > T_c$ by the change $T - T_c \to 2(T_c - T)$.

$T \ll T_c$

Here we have to take into account all the terms of expansion (13.3). It is convenient to make the shift transformation (13.20), where ρ_0 is the condensate density. We can find ρ_0 demanding S_{eff} to have a maximum after the substitution $c(p) = c^*(p) = (\rho_0\beta V)^{1/2}\delta_{p,0}$. This maximum condition can be written in the form

$$\int_0^\infty \mathrm{d}\xi \left(\frac{\tanh\frac{\beta}{2}(\xi^2 + \Delta^2)^{1/2}}{(\xi^2 + \Delta^2)^{1/2}} - \frac{\tanh\frac{\beta_c}{2}\xi}{\xi} \right) = 0, \qquad (13.26)$$

where $\Delta = Z\rho_0^{1/2}$. For instance, if $T = 0$ ($\beta = \infty$), (13.26) implies

$$\Delta(0) = \frac{\pi}{\gamma} T_c = 2c_F k_F \exp\left(-\frac{2\pi^2 c_F}{|f| Z^2 k_F^2}\right). \qquad (13.27)$$

The Bose spectrum (in the first approximation) is defined by (13.23), where $A(p)$ and $B(p)$ are the coefficients of the quadratic form in the variable $b(p), b^*(p)$. They are defined by the following formulae:

$$A(p) = (\beta V)^{-1} \sum_{p_1} G_+(p_1) G_-(p + p_1) - f_0^{-1},$$

$$B(p) = (\beta V)^{-1} \sum_{p} G_1(p_1) G_1(p + p_1), \qquad (13.28)$$

where

$$G_\pm = Z \frac{-i\omega \pm \xi(\mathbf{k})}{\omega^2 + \xi^2(\mathbf{k}) + \Delta^2},$$

$$G_1 = \frac{Z\Delta}{\omega^2 + \xi^2(\mathbf{k}) + \Delta^2}. \qquad (13.29)$$

We can split (13.23) into two equations, $A - B = 0$ and $A + B = 0$, as we did for the case $|\Delta T| \ll T_c$. The first of them ($A - B = 0$) has a solution $p = 0$ owing to the equation determining the gap:

$$A(0) - B(0) = (\beta V)^{-1} \sum_p (G_+(p) G_-(p) - G_1^2(p)) - f_0^{-1} = 0. \qquad (13.30)$$

So we can rewrite the equation $A(p) - B(p) = 0$ in the form

$$A(p) - B(p) - A(0) + B(0) = 0, \qquad (13.31)$$

or

$$\frac{Z^2}{\beta V} \sum_{p_1 + p_2 = p} \left[\frac{1}{\omega_1^2 + \xi_1^2 + \Delta^2} - \frac{(i\omega_1 + \xi_1)(i\omega_2 + \xi_2) + \Delta^2}{(\omega_1^2 + \xi_1^2 + \Delta^2)(\omega_2^2 + \xi_2^2 + \Delta^2)} \right] = 0, \qquad (13.32)$$

where $\xi_i = \xi(\mathbf{k}_i)$.

For $T \to 0$ the sum over $\mathbf{k}, \omega$ can be replaced by the integral and then we can introduce the integral over the neighbourhood of the Fermi sphere according to the rule

$$(\beta V)^{-1} \sum_p \to (2\pi)^{-4} \int d\omega\, d^3k \to$$

$$(2\pi)^{-4} \int d\omega k^2\, dk\, d\Omega \to \frac{k_F^2}{(2\pi)^2 c_F} \int d\omega\, d\xi\, d\Omega,$$

where $d\Omega$ is an element of a solid angle. In order to evaluate the integral with respect to ω_1, ξ_1, it is useful to apply the Feynman trick standard in relativistic quantum theory and based on the identity

$$(ab)^{-1} = \int_0^1 d\alpha [\alpha a + (1-\alpha)b]^{-2}.$$

We substitute $a = \omega_1^2 + \xi_1^2 + \Delta^2$, $b = \omega_2^2 + \xi_2^2 + \Delta^2$ and make replacements

$$\omega_1 \to \omega_1 + \alpha\omega, \quad \omega_2 \to -\omega_1 + (1-\alpha)\omega,$$

$$\xi_1 \to \xi_1 + \alpha c_F(\mathbf{nk}), \quad \xi_2 \to \xi_1 + (1-\alpha)c_F(\mathbf{nk}),$$

where c_F is a velocity on the Fermi surface, $\mathbf{n}$ is a unit vector orthogonal to the Fermi surface and $(\mathbf{k}, \omega) = p$, an external four-momentum. After these substitutions the integral with respect to ω_1, ξ_1 can easily be evaluated and the left-hand side of (12.31) takes the form

$$\frac{\pi k_F^2 Z^2}{(2\pi)^4 c_F} \int_0^1 d\alpha \int d\Omega \Bigg[\ln(1 + \Delta^{-2}\alpha(1-\alpha)(\omega^2 + c_F^2(\mathbf{nk})^2)) + \frac{2\alpha(1-\alpha)(\omega^2 + c_F^2(\mathbf{nk})^2)}{\Delta^2 + \alpha(1-\alpha)(\omega^2 + c_F^2(\mathbf{nk})^2)} \Bigg]. \tag{13.33}$$

For small ω, k we can expand the integrand function up to the second power of $q^2 = \omega^2 + c_F^2(\mathbf{nk})^2$ and obtain

$$\frac{k_F^2 Z^2}{8\pi^2 c_F \Delta^2} \left[\omega^2 + \frac{1}{3} c_F^2 k^2 - \frac{1}{6\Delta^2}(\omega^2 + \tfrac{2}{3}\omega^2 c_F^2 k^2 + \tfrac{1}{5} c_F^4 k^4) \right]. \tag{13.34}$$

Replacing $i\omega$ by E and putting it equal to zero we find the spectrum

$$E = uk(1 - \gamma k^2), \quad u = \frac{c_F}{3^{1/2}}, \quad \gamma = \frac{c_F^2}{45\Delta^2}. \tag{13.35}$$

This branch of the spectrum (the Bogoliubov sound) is linear in k for small k. Positivity of the coefficient γ ('dispersion') means a stability of an excitation with respect to a decay into two (or more) excitations of the same type.

Now we consider excitations described by the equation $A(p) + B(p) = 0$. Let us write the equation in the form

$$\frac{Z^2}{\beta V} \sum_{p_1 + p_2 = p} \left[\frac{(i\omega_1 - \xi_1)(i\omega_2 + \xi_2) + \Delta^2}{(\omega_1^2 + \xi_1^2 + \Delta^2)(\omega_2^2 + \xi_2^2 + \Delta^2)} + \frac{1}{\omega_1^2 + \xi_1^2 + \Delta^2} \right] = 0. \tag{13.36}$$

By using the Feynman method and by integrating with respect to ω_1, ξ_1 we write the left-hand side of (13.36) in the form

$$\frac{\pi k_F^2 Z^2}{(2\pi)^4 c_F}\int_0^1 d\alpha \int d\Omega[2 + \ln(1 + \Delta^{-2}\alpha(1-\alpha)(\omega^2 + c_F^2(\mathbf{nk})^2))]$$

$$= \frac{\pi k_F^2 Z^2}{2(2\pi)^4 c_F}\int d\Omega(\omega^2 + 4\Delta^2 + c_F^2(\mathbf{nk})^2)$$

$$\times \int_0^1 d\alpha[\Delta^2 + \alpha(1-\alpha)(\omega^2 + c_F^2(\mathbf{nk})^2)]^{-1}. \tag{13.37}$$

Expression (13.37) is equal to zero when $\omega^2 \to -4\Delta^2$ and $k \to 0$. For $E^2 = -\omega^2$ near to $4\Delta^2$ and small k the internal integral in (13.37) is reciprocal to the root $(4\Delta^2 - E^2 + c_F^2(\mathbf{nk})^2)^{1/2}$. As a result (13.36) goes to

$$\int_{-1}^{1} dx(x^2 + z^2)^{1/2} = 0 \tag{13.38}$$

or

$$(1 + z^2)^{1/2} + \frac{z^2}{2}\ln\frac{(1+z^2)^{1/2} + 1}{(1+z^2)^{1/2} - 1} = 0, \tag{13.39}$$

where

$$z^2 = \frac{(4\Delta^2 - E^2)}{c_F^2 k^2}. \tag{13.40}$$

By introducing the new variable

$$t = \ln\frac{(1+z^2)^{1/2} + 1}{(1+z^2)^{1/2} - 1} \tag{13.41}$$

we can reduce the solution of equation (13.39) to finding nontrivial ($t \neq 0$) roots of the equation

$$\sinh t + t = 0. \tag{13.42}$$

The roots group to 'quartets' $\pm a \pm bi$. For the smallest nontrivial root with respect to its modulus in the first quadrant t_1 and for asymptotics t_n for $n \to \infty$ we have

$$t_1 \approx 2.251 + i4.212,$$

$$t_n = \ln \pi(4n-1) + i\left(2\pi n - \frac{\pi}{2}\right) + o(1). \tag{13.43}$$

The sequence t_n determines the sequence of the roots of the equation

$A(p) + B(p) = 0$

$$E_n = 2\Delta - \frac{c_F^2 k^2}{4\Delta \sinh^2(t_n/2)}. \tag{13.44}$$

These roots lie on different sheets of a Riemann surface and concentrate at $E = 2\Delta$ for $n \to \infty$. The immediate physical meaning has root E_1, the first one appearing when analytic continuation from the upper to the lower half plane is performed. It corresponds to oscillations of the density which may be excited by acting on the system with the frequency near to 2Δ.

We found the Bose spectrum of a superfluid Fermi gas in the approximation corresponding to noninteracting quasiparticles of the fields c, c^*. Nevertheless, in general, the spectrum turns out to be complex.

In order to take into account interaction between quasiparticles and, in particular, to construct the kinetic theory, the higher terms in the expansion S_{eff} of fields b, b^* (fluctuations of c, c^* around their condensate values) are necessary. The summation of diagrams which reduces to solution of the kinetic equation must lead to the branch of the second sound for $0 < T < T_c$. For $T \to 0$ its velocity is

$$u_2 = \frac{c_F}{3}. \tag{13.45}$$

Let us still note that in order to build up a successive perturbation theory it is more suitable to use density-phase variables instead of c, c^* according to the formulae

$$c(\mathbf{x}, \tau) = (\rho(\mathbf{x}, \tau))^{1/2} e^{i\varphi(\mathbf{x},\tau)},$$
$$c^*(\mathbf{x}, \tau) = (\rho(\mathbf{x}, \tau))^{1/2} e^{-i\varphi(\mathbf{x},\tau)}. \tag{13.46}$$

In the new variables the branch of the Bogoliubov sound (13.35) corresponds to oscillations of the phase and excitations (13.44) correspond to oscillations of the density.

Finally, we shall make a comment on the system with a long-range Coulomb interaction which has the effective action (12.23). One can show that the Bose spectrum can be obtained from the equation

$$\det K = 0,$$
$$K = \begin{pmatrix} A(p), & B(p), & D(p) \\ B(p), & A(-p), & D(-p) \\ D(p), & D(-p), & C(p) \end{pmatrix}. \tag{13.47}$$

Here $A(p)$ and $B(p)$ are defined by (13.28) and $C(p)$ and $D(p)$ are given by functions

$$C(p) = \frac{k^2}{4\pi} + \frac{e^2}{\beta V}\sum_{p_1}(2G_1(p_1)G_1(p+p_1)$$
$$- G_+(p_1)G_+(p+p_1) - G_-(p_1)G_-(p+p_1)),$$
$$D(p) = \frac{ie}{\beta V}\sum_{p_1}(G_+(p_1)G_1(p+p_1) - G_1(p_1)G_-(p+p_1)). \tag{13.48}$$

$A(p)$, $B(p)$ and $C(p)$ are even functions of p, $D(p)$ is an odd functions of p. We can split (13.46) into two equations:

$$(A-B)C - 2D^2 = 0, \quad A + B = 0. \tag{13.49}$$

The second equation is the same as for the system without Coulomb interaction. It was investigated above. The analysis of the first equation in (13.49) shows that this equation has no sound mode solution. The phonon Bogoliubov spectrum turns into the plasma oscillation spectrum.

14

Superfluid phases in ^{3}He

The most complex and interesting of the superfluid systems is without any doubt that of ^{3}He. For the temperatures of the order 10^{-3} K there might exist several superfluid phases in ^{3}He which extend one into another on changing the external conditions – temperature, pressure and magnetic field. The difficulty in constructing a complete microscopic theory of ^{3}He makes it expedient to study simplified models analogous to that of the Bose gas for ^{4}He.

In this book we present the functional integral approach to the description of collective excitations in ^{3}He model suggested by Alonso & Popov (1977) and developed by Brusov & Popov (1980*a–c*, 1981, 1982*a*, *b*, 1984).

A simplified model of ^{3}He was discussed in the functional integral formalism in section 12. To describe collective Bose excitations in the system we move on from the integral over Fermi fields to the integral over the auxiliary Bose field which corresponds to the Bose excitations. This method was demonstrated in the previous section for the model of the superfluid Fermi gas with a pairing in the *s*-state.

The characteristic features of ^{3}He is the pairing in the *p*-state. This necessarily leads not to a scalar but to a (3×3) matrix wave function of the superfluid state. It is just this fact that guarantees the possibility of the existence of a few superfluid phases and a rich spectrum of collective excitations. In the simplified model it is enough to use a local Bose field describing the collective excitations (fields c_{ia}, c_{ia}^* introduced in section 12). In general, it is necessary to introduce a bilocal formalism, i.e. to introduce fields depending not on one but on two space-time arguments.

The ^{3}He model is defined by the effective action functional

$$S_{\text{eff}} = g_0^{-1} \sum_{p,i,a} c_{ia}^*(p)c_{ia}(p) + \tfrac{1}{2}\ln\det M(c_{ia}, c_{ia}^*)/M(0,0) \qquad (14.1)$$

(see section 12). To begin with, we consider the Landau–Ginsburg temperature region $|\Delta T| \ll T_c$. Here we can expand S_{eff} in powers of fields

c_{ia}, c^*_{ia}. Denoting

$$M(0,0) = G^{-1}, M(c_{ia}, c^*_{ia}) = G^{-1} + u, \tag{14.2}$$

we keep the first two terms ($n = 1, 2$) in the expansion

$$\tfrac{1}{2}\ln\det M(c, c^*)/M(0,0) = \tfrac{1}{2}\operatorname{Tr}\ln(I + Gu)$$
$$= -\sum_{n=1}^{\infty} \frac{1}{4n} \operatorname{Tr}(Gu)^{2n}. \tag{14.3}$$

Let us consider the quadratic form of u in S_{eff} (14.1). If $H = 0$, the form is diagonal with respect to the index a and has the form

$$- \sum_{p,i,j,a} A_{ij}(p) c^*_{ia}(p) c_{ja}(p)$$
$$= - \sum_{p,i,j,a} A_{ij}(0) c^*_{ia}(p) c_{ja}(p) - \sum_{p,i,j,a} (A_{ij}(p) - A_{ij}(0)) c^*_{ia}(p) c_{ja}(p), \tag{14.4}$$

where

$$A_{ij}(0) = -\frac{4Z^2}{\beta V} \sum_p \frac{n_i n_j}{\omega^2 + \xi^2} - g_0^{-1} \delta_{ij}$$
$$= -\delta_{ij}\left(\frac{2Z^2 k_F^2}{3\pi^2} \int_0^{c_F k_0} \frac{d\xi}{\xi} \tanh\frac{\beta\xi}{2} + g_0^{-1}\right) = -A(0)\delta_{ij}. \tag{14.5}$$

Here, when evaluating $A_{ij}(0)$ we pass to the integral over a neighbourhood of the Fermi sphere according to the rule

$$V^{-1} \sum_k \to (2\pi)^{-3} k_F^2 c_F^{-1} \int d\xi \, d\Omega.$$

Further we have

$$A_{ij}(p) - A_{ij}(0) = \frac{Z^2}{2(2\pi)^3} \int_{|k_1 - k_F| < k_0} d^3k_1 (n_{1i} - n_{2i})(n_{1j} - n_{2j})$$
$$\times \left[\frac{1}{i\omega - \xi_1 - \xi_2}\left(\tan\frac{\beta}{2}\xi_1 + \tanh\frac{\beta}{2}\xi_2\right) + \frac{1}{\xi_1}\tanh\frac{\beta}{2}\xi_1\right]. \tag{14.6}$$

For small H ($\mu H \ll T_c$) it is sufficient to take into account the contribution from H to $A_{ij}(0)$ neglecting the H-dependence of the function $A_{ij}(p) - A_{ij}(0)$ for small p. After summing up over frequencies, the following integrals arise:

$$\int_{-c_F k_0}^{c_F k_0} \frac{d\xi}{\xi \pm \mu H} \tanh\frac{\beta}{2}(\xi \pm \mu H), \quad a = 1, 2,$$

$$\int_{-c_F k_0}^{c_F k_0} \frac{d\xi}{2\xi}\left(\tanh\frac{\beta}{2}(\xi + \mu H) + \tanh\frac{\beta}{2}(\xi - \mu H)\right), \quad a = 3. \tag{14.7}$$

Only the last of them depends on H^2. If H is small, the H-dependent part of the last integral is equal to

$$\left(\frac{\beta}{2}\mu H\right)^2 \int_{-\infty}^{\infty} \frac{dx}{2x}(\tanh x)'' = -\frac{7\zeta(3)\mu^2 H^2}{2\pi^2 T^2}. \tag{14.8}$$

It is proportional to H^2. As a result the supplement to S_{eff} proportional to H^2 is equal to

$$-\frac{7\zeta(3)Z^2 k_F^2 \mu^2 H^2}{6\pi^4 T^2 c_F} \sum_{p,i} c_{i3}^*(p) c_{i3}(p). \tag{14.9}$$

We consider the fields $c_{ia}(p)$ with small three-momenta ($c_F|k| \ll T_c$) in the form of the fourth-order terms. These terms can be written as

$$-\frac{7\zeta(3)k_F^2 Z^4}{30\pi^4 T^2 c_F \beta V} \sum_{p_1+p_2=p_3+p_4} [2c_{ia}^*(p_i)c_{jb}^*(p_2)c_{ia}(p_3)c_{jb}(p_4)$$
$$+ 2c_{ia}^*(p_1)c_{ib}^*(p_2)c_{ja}(p_3)c_{jb}(p_4) + 2c_{ia}^*(p_1)c_{jb}^*(p_2)c_{ja}(p_3)c_{ib}(p_4)$$
$$- 2c_{ia}^*(p_1)c_{ja}^*(p_2)c_{ib}(p_3)c_{jb}(p_4) - c_{ia}^*(p_1)c_{ia}^*(p_2)c_{jb}(p_3)c_{jb}(p_4)]. \tag{14.10}$$

If $H = 0$ we can obtain T_c from the equation $A(0) = 0$, which is an analogue of (12.7). We can rewrite $A(0) = 0$ in the form

$$g_0^{-1} + \frac{2Z^2 k_F^2}{3\pi^2 c_F} \int_0^{c_F k_0} \frac{d\xi}{\xi} \tanh\frac{\beta}{2}\xi = 0. \tag{14.11}$$

The integral in (14.11) depends logarithmically on k_0. In order that T_c should not depend on k_0, it is necessary that g_0^{-1} depends on k_0 according to the formula

$$g_0^{-1} = g^{-1} - \frac{2Z^2 k_F^2}{3\pi^2 c_F} \ln\frac{k_0}{k_F}, \tag{14.12}$$

where g does not depend on $k_0 (g < 0)$. Substituting (14.12) and the expression

$$\int_0^{c_F k_0} \frac{d\xi}{\xi} \tanh\frac{\beta\xi}{2} \approx \ln\frac{2\beta c_F k_0}{\pi} + C \tag{14.13}$$

into (14.11), we find the temperature of the phase transition

$$T_c = \frac{2\gamma}{\pi} c_F k_F \exp\left(-\frac{3\pi^2 c_F}{2|g|Z^2 k_F^2} \right), \tag{14.14}$$

where $\ln \gamma = C$ is the Euler constant.

In the Landau–Ginsburg region $|\Delta T| \ll T_c$ we have

$$A(0) = \frac{2Z^2 k_F^2}{3\pi^2 c_F} \int_0^\infty \frac{d\xi}{\xi} \left(\tanh\frac{\beta}{2}\xi - \tanh\frac{\beta_c}{2}\xi \right)$$
$$= \frac{2Z^2 k_F^2}{3\pi^2 c_F} \ln\frac{T_c}{T} \approx \frac{2Z^2 k_F^2}{3\pi^2 c_F} \frac{T_c - T}{T_c}. \tag{14.15}$$

In order to obtain the condensate density for $T < T_c$ we make the substitutions

$$c_{i\alpha}(p) \to (\beta V)^{1/2} \frac{\pi}{Z} \left(\frac{10 T_c |\Delta T|}{7\zeta(3)} \right)^{1/2} \delta_{p0} a_{i\alpha},$$
$$c^*_{i\alpha}(p) \to (\beta V)^{1/2} \frac{\pi}{Z} \left(\frac{10 T_c |\Delta T|}{7\zeta(3)} \right) \delta_{p0} a^*_{i\alpha} \tag{14.16}$$

into (14.4) + (14.9) + (14.10). We obtain

$$-\frac{20 k_F^2 (\Delta T)^2 \beta V}{21\zeta(3) c_F} \Pi, \tag{14.17}$$

where Π depends on the matrix A with elements $a_{i\alpha}$ and on its hermitian conjugate A^+, transposed A^T and its complex conjugate A^*:

$$\Pi = -\operatorname{tr} AA^+ + \nu \operatorname{tr} A^+ A P_3 + (\operatorname{tr} AA^+)^2$$
$$+ \operatorname{tr} AA^+ AA^+ + \operatorname{tr} AA^+ A^* A^T$$
$$- \operatorname{tr} AA^T A^* A^+ - \tfrac{1}{2} \operatorname{tr} AA^T \operatorname{tr} A^+ A^*. \tag{14.18}$$

Here

$$P_3 = \begin{pmatrix} 0 & 0 & 0 \\ 0 & 0 & 0 \\ 0 & 0 & 1 \end{pmatrix} \tag{14.19}$$

is a projection operator on the third axis which is parallel to the magnetic field,

$$\nu = \frac{7\zeta(3)\mu^2 H^2}{4\pi^2 T_c \Delta T}. \tag{14.20}$$

Notice the invariance of Π under the transformation

$$A \to e^{i\alpha} UAV, \tag{14.21}$$

where α is a real parameter, U is a real orthogonal matrix of the third order, and V is a matrix of the form

$$V = \begin{pmatrix} U_2 & 0 \\ 0 & 1 \end{pmatrix}, \tag{14.22}$$

where U_2 is an orthogonal matrix of the second order.

Now we can try to find A from the minimum condition for Π:

$$\begin{aligned} \delta\Pi = &-A + \nu A P_3 + 2A\,\mathrm{tr}(AA^+) + 2AA^+A \\ &+ 2A^*A^TA - 2AA^TA^* - A^*\,\mathrm{tr}(AA^T) = 0. \end{aligned} \tag{14.23}$$

There exist several solutions of (14.23) which correspond to several superfluid phases. It turns out that only three of them may be (meta)stable, i.e. $\delta^2\Pi \geqslant 0$. These are

$$A_1 = c_1' P_3 + c_1'' P_{1,2}, \quad A_2 = c_2 P_{1,2},$$
$$A_3 = c_3 \begin{pmatrix} 1 & 0 & 0 \\ i & 0 & 0 \\ 0 & 0 & 0 \end{pmatrix}. \tag{14.24}$$

Here $P_{1,2}$ is a projection operator on two-dimensional space orthogonal to P_3. For the coefficients c_1', c_1'', c_2, c_3 we have the following formulae

$$\begin{aligned} &|c_1'|^2 = \tfrac{1}{5}(1 - 2\nu), \quad |c_1''|^2 = \tfrac{1}{10}(2 + \nu), \\ &|c_2|^2 = |c_3|^2 = \tfrac{1}{4}, \end{aligned} \tag{14.25}$$

where ν is defined by (14.20). The values of Π corresponding to A_1, A_2, A_3, are

$$\Pi_1 = -\tfrac{1}{10}(3 - 2\nu + \nu^2), \quad \Pi_2 = \Pi_3 = -\tfrac{1}{4}. \tag{14.26}$$

If $\nu = 0$ ($H = 0$), Π is minimal ($\Pi = -0.3$) for $A = A_1 = cI$. It is the so-called Balian–Werthamer superfluid phase (or B-phase). Two other phases are the A_2 (planar or 2D-phase) and the A_3 (Anderson–Morel–Brinkman or A-phase). Π is equal to $-1/4$ for both A- and 2D-phases.

The investigation of the second variation form $\delta^2\Pi$ shows that $\delta^2\Pi \geqslant 0$ for both B- and A-phases at $\nu = 0$, and $\delta^2\Pi \gtrless 0$ for the 2D-phase at $\nu = 0$. The notation $\delta^2\Pi \gtrless 0$ means that this quadratic form has both positive and negative eigenvalues. These results show that the A- and B-phases are (meta)stable, whereas the 2D-phase is unstable in zero magnetic field ($H = 0$).

If we now begin to switch on the magnetic field, the B-phase becomes 'deformed', and turns into the 2D-phase, if $\nu = 1/2$. Moreover, the solution A_1 becomes meaningless if $\nu > 1/2$ (see the first of formulae (14.25)). So we can say that there is a phase transition of the B-phase into the 2D-phase at

$$\nu = 1/2, \quad \text{or} \quad H_c^2 = 2\pi^2 T_c \Delta T / 7\zeta(3)\mu^2. \tag{14.27}$$

This conclusion can be supported by the calculation of $\delta^2\Pi$. (See formulae (14.29) below.) If $\nu < 1/2$, we have $\delta^2\Pi_1 \geqslant 0$, $\delta^2\Pi_2 \gtrless 0$. If $\nu > 1/2$, $\delta^2\Pi_2 \geqslant 0$, and the solution A_1 does not exist at all. Besides the second-order transition from the B- to the 2D-phase, there is a competitive first-order transition from the B- to the A-phase for the same value $\nu = 1/2$ (since the energies of the A- and 2D-phases are, in the first approximation, the same, and the A-phase is metastable, i.e. $\delta^2\Pi_3 \geqslant 0$ for all ν). In the real ^{3}He there exists only the first-order phase transition from the B- into the A-phase.

Now we investigate the stability of various superfluid phases under small fluctuations. The necessary stability condition is a non-negativity of the second variation form $\delta^2\Pi$. We present the expressions $\delta^2\Pi_i$ for B-, 2D- and A-phases:

$$\begin{aligned}
\delta^2\Pi_1 = {} & \frac{\nu+2}{5}[3u_{11}^2 + 3u_{22}^2 + 2u_{11}u_{22} + (u_{12}+u_{21})^2 + u_{13}^2 + u_{23}^2] \\
& + \frac{2(1-2\nu)}{5}(u_{31}^2 + u_{32}^2 + 3u_{33}^2) \\
& + \frac{4}{5}\left[\frac{(1-2\nu)(2+\nu)}{2}\right]^{1/2}(u_{11}u_{33} + u_{22}u_{33} + u_{13}u_{31} + u_{23}u_{32}) \\
& + \frac{4-3\nu}{5}(v_{11}^2 + v_{22}^2) + \frac{8-\nu}{5}(v_{12}^2 + v_{21}^2) \\
& + \frac{2(2+\nu)}{5}(v_{33}^2 - v_{11}v_{22} - v_{12}v_{21}) + \frac{8+9\nu}{5}(v_{13}^2 + v_{23}^2) \\
& + \frac{8(1-2\nu)}{5}(v_{31}^2 + v_{32}^2) \\
& - \frac{4}{5}\left[\frac{(1-2\nu)(2+\nu)}{2}\right]^{1/2}(v_{11}v_{33} + v_{22}v_{33} + v_{13}v_{31} + v_{23}v_{32}),
\end{aligned}$$

$$\begin{aligned}
\delta^2\Pi_2 = {} & (\nu - \tfrac{1}{2})u_{33}^2 + (\nu + \tfrac{1}{2})v_{33}^2 + \nu(u_{13}^2 + u_{23}^2) \\
& + (\nu + 2)(v_{13}^2 + v_{23}^2) + \tfrac{1}{2}[3u_{11}^2 + 3u_{22}^2 + 2u_{12}u_{22} + (u_{12} + u_{21})^2] \\
& + \tfrac{1}{2}[3v_{11}^2 + 3v_{21}^2 - 2v_{12}v_{21} + (v_{11} - v_{22})^2],
\end{aligned}$$

$$\begin{aligned}\delta^2\Pi_3 = {} & \nu[u_{13}^2 + u_{23}^2 + u_{33}^2 + v_{13}^2 + v_{23}^2 + v_{33}^2] \\ & + (u_{11} + v_{21})^2 + (u_{22} - v_{12})^2 + (u_{23} - v_{13})^2 \\ & + \tfrac{1}{2}[(u_{11} - v_{21})^2 + (u_{21} + v_{11})^2 + (u_{12} - v_{22})^2 \\ & + (u_{22} + v_{12})^2 + (u_{13} - v_{23})^2 + (u_{23} + v_{13})^2].\end{aligned} \tag{14.28}$$

Here $u_{ia} = \mathrm{Re}\,\delta a_{ia}$, $v_{ia} = \mathrm{Im}\,\delta a_{ia}$.

For $\nu < 1/2$ the variation $\delta^2\Pi_1$ is non-negative and for $\nu > 1/2$ that of $\delta^2\Pi_2$ is non-negative. The variation $\delta^2\Pi_3$ is non-negative for all ν. Thus, the B-phase must pass either into the 2D-phase or into the A-phase for sufficiently large ν.

Besides solutions (14.24) corresponding to the B-, 2D- and A-phases, (14.23) has other solutions, for instance

$$A_4 = c_4 P_3, \quad A_5 = c_5 C_5, \quad A_6 = c_6' C_3 + c_6'' C_5, \quad A_7 = c_7 C_7, \tag{14.29}$$

where P_3 is the projection operator in (14.19), and

$$C_5 = \begin{pmatrix} 0 & 0 & 0 \\ 0 & 0 & 0 \\ 1 & \mathrm{i} & 0 \end{pmatrix}, \quad C_7 = \begin{pmatrix} 1 & \mathrm{i} & 0 \\ \mathrm{i} & -1 & 0 \\ 0 & 0 & 0 \end{pmatrix}, \tag{14.30}$$

$$|c_4|^2 = \tfrac{1}{3}(1-\nu), \quad |c_5|^2 = \tfrac{1}{12}, \quad |c_6'|^2 = \tfrac{1}{8}(2 - 3\nu),$$

$$|c_6''|^2 = \frac{\nu}{8}, \quad |c_7|^2 = \tfrac{1}{16}. \tag{14.31}$$

The corresponding values of Π are

$$\begin{aligned} & \Pi_4 = -\tfrac{1}{6}(1-\nu)^2, \quad \Pi_5 = -\tfrac{1}{12}, \\ & \Pi_6 = -\tfrac{1}{8}(2 - 4\nu + 3\nu^2), \quad \Pi_7 = -\tfrac{1}{8}. \end{aligned} \tag{14.32}$$

These four phases A_4, A_5, A_6 and A_7 are energetically disadvantageous compared with the B-, A- and 2D-phases. Moreover, they are unstable under small fluctuations because $\delta^2\Pi \gtrless 0$ for all these phases.

Thus, in the model system in question, the stability condition $\delta^2\Pi \geqslant 0$ is satisfied only for the A-phase, the B-phase ($H \leqslant H_c$) and the 2D-phase ($H \geqslant H_c$).

The quadratic form $\delta^2\Pi_1$ of 18 variables u_{ia}, v_{ia} (B-phase, $\nu < 1/2$) has four eigenvectors corresponding to zero eigenvalues. The form $\delta^2\Pi_2$ (2D-phase, $\nu > 1/2$) has six zero eigenvectors and $\delta^2\Pi_3$ (A-phase) has nine eigenvectors for $\nu = 0$. This is connected with the existence of four phonon type branches of the Bose spectrum in the B-phase, of six such branches in the 2D-phase and of nine in the A-phase (for $H = 0$).

15

Collective excitations in the B-phase of ^{3}He

Let us first consider the Bose spectrum of the system for $|\Delta T| \ll T_c$. For $T > T_c$ all branches of the spectrum are pure imaginary and $|E(\mathbf{k})| \ll T_c$. By analytic continuation of (14.6) on the domain $|\omega| \ll T$ we obtain

$$A_{ij}(p) - A_{ij}(0) = \frac{Z^2 k_F^2}{12\pi T c_F}\left[\omega\delta_{ij} + \frac{7\zeta(3)c_F^2}{10\pi^3 T}(k^2\delta_{ij} + k_i k_j)\right] \tag{15.1}$$

for $\omega > 0$. By taking into account the value of $A(0)$ given by (14.15) and the magnetic supplement (14.9), we obtain the following branches of the spectrum for $T > T_c$:

$$\begin{aligned} E_{a,\parallel}(\mathbf{k}) &= -\mathrm{i}\left(\frac{21\zeta(3)c_F^2 k^2}{10\pi^3 T_c} + \frac{8}{\pi}(T - T_c) + \delta_{a3}\frac{14\zeta(3)\mu^2 H^2}{\pi^3 T_c}\right), \\ E_{a,\perp}(\mathbf{k}) &= -\mathrm{i}\left(\frac{7\zeta(3)c_F^2 k^2}{10\pi^3 T_c} + \frac{8}{\pi}(T - T_c) + \delta_{a3}\frac{14\zeta(3)\mu^2 H^2}{\pi^3 T_c}\right). \end{aligned} \tag{15.2}$$

Here a is a spinorial index of the corresponding branch and the subscripts $\parallel$ and $\perp$ indicate that the vector index is 'parallel' or 'perpendicular' to the direction of the wave vector $\mathbf{k}$.

For $T < T_c$ it is necessary to take into account the form of the fourth degree. It is not difficult to obtain the Bose spectrum after the shift transformation $c_{ia}(p) \to c_{ia}(p) + c_{ia}^{(0)}(p)$. All its branches have the form

$$E(\mathbf{k}) = -\mathrm{i}\alpha k^2 - \mathrm{i}\Gamma, \tag{15.3}$$

$\Gamma \geqslant 0$ with $\Gamma = 0$ for four branches in the B-phase, six branches in the 2D-phase and nine in the A-phase. It is only these branches that go into the branches of the phonon spectrum for decreasing temperature.

Let us now go on to the temperature domain $T_c - T \sim T_c$. In this domain we expand ln det in (14.1) in powers of the deviation $c_{ia}(p)$ from the condensate quantity $c_{ia}^{(0)}(p)$ (different for different phases). We make the shift transformation

$$c_{ia}(p) = c_{ia}^{(0)}(p) + a_{ia}(p) \tag{15.4}$$

and extract the quadratic form of a_{ia}, a^*_{ia} from S_{eff}

$$-\sum_p a^*_{ia}(p)a_{jb}(p)A_{ijab}(p) - \frac{1}{2}\sum_p (a_{ia}(p)a_{jb}(-p)B_{ijab}(p) + a^*_{ia}(p)a^*_{jb}(-p)\bar{B}_{ijab}(p)). \tag{15.5}$$

This form determines the Bose spectrum in the first approximation by using the equation

$$\det Q = 0. \tag{15.6}$$

Here Q is a matrix of the quadratic form which is determined by the coefficient tensors A_{ijab} and B_{ijab} in (15.5). If $T \to 0$, the quantities A_{ijab} and B_{ijab} are proportional to the integrals of the Green functions of fermions. These integrals are more straightforwardly calculated by applying the Feynman method. With its help it is easy to perform integration over the variables ω and ξ and then over the angular variables and parameter α.

Now we are in a position to obtain all the branches of the Bose spectrum in the B-phase of a ^{3}He-type system and then also in the A-phase (section 16).

If $H = 0$, the condensate wave-function of the B-phase has the form

$$c^{(0)}_{ia}(p) = (\beta V)^{1/2}\delta_{p0}\delta_{ia}c, \tag{15.7}$$

and c can be obtained from the equation

$$\frac{3}{g_0} + \frac{4Z^2}{\beta V}\sum_p (\omega^2 + \xi^2 + 4c^2 Z^2)^{-1}$$
$$= \frac{3}{g_0} + \frac{2Z^2}{\pi^2 c_F}\int_0^{c_F k_0} \frac{d\xi}{(\xi^2 + 4c^2Z^2)^{1/2}} \tanh\frac{\beta}{2}(\xi^2 + 4c^2Z^2)^{1/2} = 0. \tag{15.8}$$

Using (14.11) and the notation $2cZ = \Delta$, we may write (15.8) as

$$\int_0^\infty d\xi\left((\xi^2 + \Delta^2)^{-1/2}\tanh\frac{\beta}{2}(\xi^2 + \Delta^2)^{1/2} - \xi^{-1}\tanh\frac{\beta}{2}\xi\right) = 0. \tag{15.9}$$

From (15.9) we obtain, for $\Delta_0 = \Delta(T = 0)$,

$$\Delta_0 = \frac{\pi}{\gamma}T_c = c_F k_F \exp\left(-\frac{3\pi^2 c_F}{2|g|Z^2 k_F^2}\right). \tag{15.10}$$

We also have the following formula for S_{eff}:

$$\Delta p_B = \lim_{T\to 0,\, V\to\infty} (S_{\text{eff}}/\beta V) = T_c^2 k_F^2/4\gamma^2 c_F. \tag{15.11}$$

Δp_B is the correction to the pressure of the ideal Fermi gas in the corresponding superfluid (B-)phase.

Analogous formulae can also be obtained for the A- and 2D-phases

$$\Delta p_A = \Delta p_{2D} = (\Delta p_B)\frac{e^{5/3}}{6}. \tag{15.12}$$

Formulae (15.11) and (15.12) show that at $T = 0$ the optimum phase is the symmetric B-phase, and the A- and 2D-phases have 'equal opportunities'. The same is true in the Ginsburg–Landau region $|\Delta T| \ll T_c$. All other phases (14.29) can be shown to be less preferable as compared with the B-, A- and 2D-phases. Moreover they are unstable under small fluctuations.

Let us return to the B-phase. We shall determine all the branches of the Bose spectrum $E(\mathbf{k})$. At small $\mathbf{k}$ all the branches are of the form

$$E^2(\mathbf{k}) = \Omega^2 + \alpha k^2. \tag{15.13}$$

All the branches can be obtained from the equation $\det Q = 0$ (15.6), where Q is the matrix of the quadratic form (15.5). This quadratic form varies for different superfluid phases. For the B-phase it is equal to

$$\begin{aligned}
&-\sum_p a^*_{ia}(p) a_{ja}(p) \\
&\times\left[-g_0^{-1}\delta_{ij} - \frac{4Z^2}{\beta V}\sum_{p_1+p_2=p} n_{1i}n_{1j}\varepsilon(-p_1)\varepsilon(-p_2)G(p_1)G(p_2)\right] \\
&-\frac{1}{2}\sum_p (a_{ia}(p)a_{jb}(-p) + a^*_{ia}(p)a^*_{jb}(-p)) \\
&\times\frac{4Z^2\Delta^2}{\beta V}\sum_{p_1+p_2=p} G(p_1)G(p_2)(2n_{1a}n_{1b}n_{1i}n_{1j} - \delta_{ab}n_{1i}n_{1j}), \\
&\varepsilon(p) = i\omega - \xi, \quad G(p) = (\omega^2 + \xi^2 + \Delta^2)^{-1}.
\end{aligned} \tag{15.14}$$

After the substitutions

$$\begin{aligned}
a_{ia}(p) &= u_{ia}(p) + iv_{ia}(p) \\
a^*_{ia}(p) &= u_{ia}(p) - iv_{ia}(p)
\end{aligned} \tag{15.15}$$

(15.14) breaks up into two independent forms, one of which depends on u_{ia} and the other on v_{ia}:

$$\begin{aligned}
&-\sum_p (A_{ij}u_{ia}u_{ja} + B_{ijab}u_{ia}u_{jb}) \\
&-\sum_p (A_{ij}v_{ia}v_{ja} - B_{ijab}v_{ia}v_{jb}).
\end{aligned} \tag{15.16}$$

If we go on to the integral near the Fermi sphere and use the Feynman trick we can express $A_{ij}(p)$ in the following form ($T \to 0$):

$$A_{ij}(p) = -\frac{4Z^2 k_F^2}{(2\pi)^4 c_F} \int_0^1 d\alpha \int d\omega_1 \, d\xi_1 \, d\Omega_1 n_{1i} n_{1j}$$

$$\times \left[\frac{(\xi_1 + i\omega_1)(\xi_2 + i\omega_2)}{[\alpha(\omega_1^2 + \xi_1^2 + \Delta^2) + (1-\alpha)(\omega_2^2 + \xi_2^2 + \Delta^2)]^2} - \frac{1}{\omega_1^2 + \xi_1^2 + \Delta^2} \right], \tag{15.17}$$

where the term $\delta_{ij} g_0^{-1}$ in $A_{ij}(p)$ is eliminated by using the identity

$$\delta_{ij} g_0^{-1} + \frac{4Z^2}{\beta V} \sum_p n_i n_j (\omega^2 + \xi^2 + \Delta^2)^{-1} = 0, \tag{15.18}$$

which determines the gap Δ.

We then calculate directly the integrals with respect to ω_1 and ξ_1. We obtain the formula

$$A_{ij}(p) = \frac{Z^2 k_F^2}{4\pi^3 c_F} \int_0^1 d\alpha \int d\Omega_1 n_{1i} n_{1j}$$

$$\times \left[\ln(1 + \Delta^{-2} \alpha(1-\alpha)(\omega^2 + c_F^2(\mathbf{nk})^2)) \right.$$

$$\left. + \frac{\Delta^2 + 2\alpha(1-\alpha)(\omega^2 + c_F^2(\mathbf{nk})^2)}{\Delta^2 + \alpha(1-\alpha)(\omega^2 + c_F^2(\mathbf{nk})^2)} \right]. \tag{15.19}$$

A similar procedure for B_{ijab} yields

$$B_{ijab} = \frac{Z^2 k_F^2}{4\pi^3 c_F} \int_0^1 d\alpha \int d\Omega_1 n_{1i} n_{1j} (2 n_{1a} n_{1b} - \delta_{ab})$$

$$\times \frac{\Delta^2}{\Delta^2 + \alpha(1-\alpha)(\omega^2 + c_F^2(\mathbf{nk})^2)}. \tag{15.20}$$

First we put $\mathbf{k} = 0$ in (15.19) and (15.20). Then the integrals with respect to the angle variables and the parameter α separate and can be easily calculated as

$$\int d\Omega_1 n_{1i} n_{1j} = \frac{4\pi}{3} \delta_{ij},$$

$$\int d\Omega_1 n_{1i} n_{1j} (2 n_{1a} n_{1b} - \delta_{ab}) = \frac{4\pi}{15} (2\delta_{ai}\delta_{bj} + 2\delta_{aj}\delta_{bi} - 3\delta_{ab}\delta_{ij}),$$

$$\int_0^1 d\alpha \left[\ln(1 + \Delta^{-2}\alpha(1-\alpha)\omega^2) + \frac{\Delta^2 + 2\alpha(1-\alpha)\omega^2}{\Delta^2 + \alpha(1-\alpha)\omega^2} \right] = f(\omega), \tag{15.21}$$

$$\int_0^1 \frac{\Delta^2\, d\alpha}{\Delta^2 + \alpha(1-\alpha)\omega^2} = g(\omega), \tag{15.22}$$

where

$$f(\omega) = (\omega^2 + 2\Delta^2)h(\omega), \quad g(\omega) = 2\Delta^2 h(\omega),$$
$$h(\omega) = [\omega(\omega^2 + 4\Delta^2)^{1/2}]^{-1} \ln \frac{(\omega^2 + 4\Delta^2)^{1/2} + \omega}{(\omega^2 + 4\Delta^2)^{1/2} - \omega}. \tag{15.23}$$

Substituting (15.21)–(15.23) in (15.19) and (15.20) we obtain the expressions for $A_{ij}(p)$, $B_{ijab}(p)$ at $k = 0$:

$$A_{ij}(p) = \frac{Z^2 k_F^2}{3\pi^2 c_F} \delta_{ij} f(\omega),$$

$$B_{ijab}(p) = \frac{Z^2 k_F^2}{15\pi^2 c_F} (2\delta_{ai}\delta_{bj} + 2\delta_{aj}\delta_{bi} - 3\delta_{ab}\delta_{ij}) g(\omega). \tag{15.24}$$

Substituting these in (15.16), we come to the quadratic form

$$-\frac{Z^2 k_F^2}{15\pi^2 c_F} \sum_p \{5f(\omega)(u_{ia}u_{ia} + v_{ia}v_{ia}) + g(\omega)[2(u_{ia}u_{ai} - v_{ia}v_{ai})$$
$$+ 2(u_{aa}u_{ii} - v_{aa}v_{ii}) - 3(u_{ia}u_{ia} - v_{ia}v_{ia})]\}. \tag{15.25}$$

From this expression we easily obtain the frequencies of all 18 branches.

The expression under the sign of summation with respect to p breaks into several independent forms containing the variables

$$(u_{12}, u_{21}), \quad (u_{23}, u_{32}), \quad (u_{31}, u_{13}), \quad (u_{11}, u_{22}, u_{33}),$$
$$(v_{12}, v_{21}), \quad (v_{23}, v_{32}), \quad (v_{31}, v_{13}), \quad (v_{11}, v_{22}, v_{33}). \tag{15.26}$$

The form of the variables (u_{12}, u_{21}) is given by

$$[5f(\omega) - 3g(\omega)](u_{12}^2 + u_{21}^2) + 4g(\omega)u_{12}u_{21}. \tag{15.27}$$

Equating its determinant to zero, we get

$$5[f(\omega) - g(\omega)][5f(\omega) - g(\omega)] = 0. \tag{15.28}$$

If the first factor in (15.28) vanishes to zero, we arrive at the equation $\omega^2 h(\omega) = 0$, which yields the branch $E^2 = 0$. This is one of the phonon branches, equal to zero at $\mathbf{k} = 0$. The vanishing of the second factor in (15.28) leads to the equation $(5\omega^2 + 8\Delta^2)h(\omega) = 0$ and yields the branch $E^2 = 8\Delta^2/5$. The forms of the variables (u_{23}, u_{32}), (u_{31}, u_{13}) have at $\mathbf{k} = 0$ the same coefficients as (15.26). They yield two more branches with $E^2 = 0$ and two further branches with $E^2 = 8\Delta^2/5$.

We consider now the form of (u_{11}, u_{22}, u_{33})

$$[5f(\omega)+g(\omega)](u_{11}^2+u_{22}^2+u_{33}^2)+4g(\omega)(u_{11}u_{22}+u_{22}u_{33}+u_{33}u_{11}). \tag{15.29}$$

Equating its determinant to zero we obtain

$$[5f(\omega)-g(\omega)]^2 5[f(\omega)+g(\omega)]=0. \tag{15.30}$$

The equation $(5f-g)^2=0$ gives two more branches with $E^2 = 8\Delta^2/5$, while $f+g=0$ is equivalent to the equation $(\omega^2+4\Delta^2)h(\omega)=0$ and gives the branch $E^2=4\Delta^2$. Thus, the variables u_{ia} give five branches $E^2 = 8\Delta^2/5$, three branches $E^2=0$ and one branch $E^2=4\Delta^2$.

We examine now the v branches. The forms of the variables (v_{12}, v_{21}), (v_{23}, v_{32}) and (v_{31}, v_{13}) are the same at $\mathbf{k}=0$. The coefficients of these forms differ from the corresponding coefficients of the forms by the substitution $g(\omega) \to -g(\omega)$. Therefore the counterpart of (15.28) is

$$5[f(\omega)+g(\omega)][5f(\omega)+g(\omega)]=0. \tag{15.31}$$

The equality $f+g=0$ yields three branches $E^2=4\Delta^2$ (one for each of the three forms), while $5f+g=0$ is equivalent to $(5\omega^2+12\Delta^2)h(\omega)=0$ and yields the three branches with $E^2=12\Delta^2/5$.

The equation corresponding to the form containing (v_{11}, v_{22}, v_{33}) is obtained from (15.30) by making a substitution $g(\omega)\to -g(\omega)$ and it reads

$$[5f(\omega)+g(\omega)]^2 5[f(\omega)-g(\omega)]=0. \tag{15.32}$$

The equation $(5f+g)^2=0$ gives two more branches with $E^2=12\Delta^2/5$, and $f-g=0$ gives $E^2=0$, corresponding to a phonon (acoustic) branch that vanishes at $k=0$. The variables v_{ia} thus yield five branches with $E^2=12\Delta^2/5$, three branches with $E^2=4\Delta^2$ and one branch with $E^2=0$. Here is a list of all the frequencies obtained together with their corresponding variables:

$$\begin{aligned}
&\Omega^2=0 \quad u_{12}-u_{21},\ u_{23}-u_{32},\ u_{31}-u_{13},\ v_{11}+v_{22}+v_{33};\\
&\Omega^2=4\Delta^2 \quad v_{12}-v_{21},\ v_{23}-v_{32},\ v_{31}-v_{13},\ u_{11}+u_{22}+u_{33};\\
&\Omega^2=8\Delta^2/5 \quad u_{12}+u_{21},\ u_{23}+u_{32},\ u_{31}+u_{13},\\
&\qquad\qquad u_{11}-u_{22},\ u_{11}+u_{22}-2u_{33};\\
&\Omega^2=12\Delta^2/5 \quad v_{12}+v_{21},\ v_{23}+v_{32},\ v_{31}+v_{13},\\
&\qquad\qquad v_{11}-v_{22},\ v_{11}+v_{22}-2v_{33}.
\end{aligned} \tag{15.33}$$

We note that the branches with $\Omega = 0$ and $\Omega = 2\Delta$ are 'dual' in the sense that the variables corresponding to them differ by the substitution $u_{ia} \rightleftarrows v_{ia}$. In this sense, the branches with $\Omega = \Delta(8/5)^{1/2}$ and $\Omega = \Delta(12/5)^{1/2}$ are dual. We also note that the sum of the squares of the frequencies of dual branches is equal to $4\Delta^2$.

Now we can calculate all the branches of the Bose spectrum with corrections $\sim k^2$. The most labour-consuming task here is the analysis of the branches with $\Omega = 2\Delta$, of which the coefficients are complex. Therefore, we consider first the other branches, in which the corrections to the spectrum can be obtained by expanding the coefficient tensors $A_{ij}(p)$, $B_{ijab}(p)$ in powers of k^2, confining ourselves to terms $\sim k^2$. We have

$$\int_0^1 \mathrm{d}\alpha \left[\ln(1 + \Delta^{-2}\alpha(1-\alpha)(\omega^2 + c_F^2(\mathbf{nk})^2)) + \frac{\Delta^2 + 2\alpha(1-\alpha)(\omega^2 + c_F^2(\mathbf{nk})^2)}{\Delta^2 + \alpha(1-\alpha)(\omega^2 + c_F^2(\mathbf{nk})^2)} \right]$$

$$= f(\omega) + \frac{\mathrm{d}f(\omega)}{\mathrm{d}\omega^2} c_F^2(\mathbf{nk})^2 + O(k^4),$$

$$\int_0^1 \frac{\Delta^2 \mathrm{d}\alpha}{\Delta^2 + \alpha(1-\alpha)(\omega^2 + c_F^2(\mathbf{nk})^2)} = g(\omega) + \frac{\mathrm{d}g(\omega)}{\mathrm{d}\omega^2} c_F^2(\mathbf{nk})^2 + O(k^4). \quad (15.34)$$

We can thus obtain corrections to the quadratic form (15.19). Evaluating the corresponding integrals with respect to the angle variables

$$\int \mathrm{d}\Omega_1 n_{1i} n_{1j} (\mathbf{n}_1\mathbf{k})^2 = \frac{4\pi}{15}(k^2\delta_{ij} + k_i k_j),$$

$$\int \mathrm{d}\Omega_1 (2n_{1a}n_{1b} - \delta_{ab}) n_{1i} n_{1j} (\mathbf{n}_1\mathbf{k})^2$$

$$= \frac{4\pi}{105}[k^2(2\delta_{ai}\delta_{bj} + 2\delta_{aj}\delta_{bi} - 5\delta_{ab}\delta_{ij}) + 4k_a k_b \delta_{ij}$$

$$- 10k_i k_j \delta_{ab} + 2k_a k_i \delta_{bj} + 2k_b k_j \delta_{ai} + 2k_a k_j \delta_{bi} + 2k_b k_i \delta_{aj}], \quad (15.35)$$

we write the correction to (15.25) in the form

$$-\frac{Z^2 k_F^2 c_F}{15\pi^2} \sum_p \left\{ \frac{\mathrm{d}f(\omega)}{\mathrm{d}\omega^2} [k^2(u_{ia}u_{ia} + v_{ia}v_{ia}) + 2k_i k_j (u_{ia}u_{ja} + v_{ia}v_{ja})] \right.$$

$$+ \frac{1}{7}\frac{\mathrm{d}g(\omega)}{\mathrm{d}\omega^2}[k^2((u_{aa}u_{bb} - v_{aa}v_{bb}) + 2(u_{ia}u_{ai} - v_{ia}v_{ai})$$

$$- 5(u_{ia}u_{ia} - v_{ia}v_{ia})) + k_i k_j(4(u_{ai}u_{aj} - v_{ai}v_{aj}) - 10(u_{ia}u_{ja} - v_{ia}v_{ja})$$

$$+ 8(u_{ij}u_{aa} - v_{ij}v_{aa}) + 8(u_{ia}u_{aj} - v_{ia}v_{aj}))]\Big\}. \tag{15.36}$$

Since the B-phase is isotropic and there is no preferred direction in it, it suffices to consider excitations that propagate on some fixed direction, say along the third axis. After the substitutions $k_1 = k_2 = 0$ and $k_3 = k$, the correction (15.36) breaks up into a sum of forms with the same variables (15.26) as the main form (15.25). Adding (15.36) to (15.25), we obtain (at $k_1 = k_2 = 0, k_3 = k$) under the summation symbol the sum of the following forms:

$$(w_{12}^2 + w_{21}^2)\left[5f(\omega) \mp 3g(\omega) + c_F^2 k^2\left(\frac{df}{d\omega^2} \mp \frac{5}{7}\frac{dg}{d\omega^2}\right)\right]$$

$$\pm 4w_{12}w_{21}\left(g(\omega) + \frac{1}{7}c_F^2 k^2 \frac{dg}{d\omega^2}\right),$$

$$w_{31}^2\left[5f(\omega) \mp 3g(\omega) + c_F^2 k^2\left(3\frac{df}{d\omega^2} \mp \frac{15}{7}\frac{dg}{d\omega^2}\right)\right]$$

$$+ w_{13}^2\left[5f(\omega) \mp 3g(\omega) + c_F^2 k^2\left(\frac{df}{d\omega^2} \mp \frac{1}{7}\frac{dg}{d\omega^2}\right)\right]$$

$$\pm 4w_{13}w_{31}\left(g(\omega) + \frac{3}{7}c_F^2 k^2 \frac{dg}{d\omega^2}\right),$$

$$(w_{11}^2 + w_{22}^2)\left[5f(\omega) \pm g(\omega) + c_F^2 k^2\left(\frac{df}{d\omega^2} \mp \frac{1}{7}\frac{dg}{d\omega^2}\right)\right]$$

$$+ w_{33}^2\left[5f(\omega) \pm g(\omega) + c_F^2 k^2\left(3\frac{df}{d\omega^2} \pm \frac{9}{7}\frac{dg}{d\omega^2}\right)\right]$$

$$\pm 4w_{11}w_{22}\left(g(\omega) + \frac{1}{7}c_F^2 k^2 \frac{dg}{d\omega^2}\right)$$

$$\pm 4w_{33}(w_{11} + w_{22})\left(g(\omega) + \frac{3}{7}c_F^2 k^2 \frac{dg}{d\omega^2}\right). \tag{15.37}$$

We must put here $w_{ia} = u_{ia}$ and take the upper signs $\pm, \mp$ or else put $w_{ia} = v_{ia}$ and take the lower signs. The forms with the variables w_{23}, w_{32} are obtained from the second form of (15.37) by making a replacement $(w_{13}, w_{31}) \to (w_{23}, w_{32})$.

It is easy to obtain from (15.37) all the spectrum branches in question (except $\Omega = 2\Delta$) with corrections $\sim k^2$. Consider, for example, the first of the forms (15.37). Putting $w_{12}=u_{12}, w_{21}=u_{21}$ (taking the upper signs in $\pm, \mp$) we put the determinant of the form to zero

$$5\left[f(\omega)-g(\omega)+\frac{c_F^2k^2}{5}\frac{d}{d\omega^2}(f(\omega)-g(\omega))\right]$$

$$\times\left[5f(\omega)-g(\omega)+\frac{c_F^2k^2}{5}\frac{d}{d\omega^2}(f(\omega)-\tfrac{3}{7}g(\omega))\right]=0. \qquad (15.38)$$

Equating to zero the first factor in (15.38) yields

$$\omega^2h(\omega)+\frac{c_F^2k^2}{5}\frac{d}{d\omega^2}\omega^2h(\omega)=0$$

or

$$\omega^2+\frac{c_F^2k^2}{5}+\omega^2\frac{c_F^2k^2}{5}\frac{d\ln h(\omega)}{d\omega^2}=0.$$

The last term here is of higher order than the others, since $d\ln h(\omega)/d\omega^2$ is finite as $\omega\to 0$. As a result, we obtain the phonon branch of the longitudinal spin oscillations

$$E^2=\tfrac{1}{5}c_F^2k^2. \qquad (15.39)$$

All the remaining phonon branches can be obtained using the approach described here when considering the forms (u_{13}, u_{31}), (u_{23}, u_{32}), (v_{11}, v_{22}, v_{33}).

Equating to zero the second factor in (15.38) we obtain the equation

$$(5\omega^2+8\Delta^2)h(\omega)+\frac{c_F^2k^2}{7}\frac{d}{d\omega^2}(7\omega^2+8\Delta^2)h(\omega)=0,$$

which yields the spectrum branch

$$E^2=\frac{8\Delta^2}{5}+\frac{c_F^2k^2}{5}\left(1-\frac{16\Delta^2}{35}\frac{d\ln h(\omega)}{d\omega^2}\right)\Bigg|_{\omega^2=-8\Delta^2/5}$$

Using the formula

$$\frac{d\ln h(\omega)}{d\omega^2}=-\frac{1}{2\omega^2}-\frac{1}{2(\omega^2+4\Delta^2)}$$

$$-\left[\omega(\omega^2+4\Delta^2)^{1/2}\ln\frac{(\omega^2+4\Delta^2)^{1/2}+\omega}{(\omega^2+4\Delta^2)^{1/2}-\omega}\right]^{-1}$$

and substituting $\omega=i\Delta(8/5)^{1/2}$, we obtain

$$\frac{d\ln h(\omega)}{d\omega^2}=\frac{5}{48\Delta^2}\left(1-\frac{2(6^{1/2})}{\arctan 2(6^{1/2})}\right).$$

This leads to the spectrum branch corresponding to the variable $u_{12}+u_{21}$

$$E^2=\frac{8\Delta^2}{5}+\frac{c_F^2k^2}{105}\left[20+\frac{2(6^{1/2})}{\arctan 2(6^{1/2})}\right]. \tag{15.40}$$

Similar calculations are easily made for the remaining branches with the exception of those with $\Omega=2\Delta$.

The branches of the Bose spectrum together with their corresponding variables are the four phonon branches:

$$\begin{aligned} &E^2=c_F^2k^2/5, \quad u_{12}-u_{21};\\ &E^2=2c_F^2k^2/5, \quad u_{13}-u_{31}, u_{23}-u_{23};\\ &E^2=c_F^2k^2/3, \quad v_{11}+v_{22}+v_{33}; \end{aligned} \tag{15.41}$$

five u branches with $\Omega=\Delta(8/5)^{1/2}$:

$$\begin{aligned} &E^2=\frac{8\Delta^2}{5}+\frac{c_F^2k^2}{105}\left[20+\frac{2(6^{1/2})}{\arctan 2(6^{1/2})}\right], u_{12}+u_{21}, u_{11}-u_{22};\\ &E^2=\frac{8\Delta^2}{5}+\frac{c_F^2k^2}{210}\left[85-\frac{2(6^{1/2})}{\arctan 2(6^{1/2})}\right], u_{13}+u_{31}, u_{23}+u_{32};\\ &E^2=\frac{8\Delta^2}{5}+\frac{c_F^2k^2}{105}\left[50-\frac{2(6^{1/2})}{\arctan 2(6^{1/2})}\right], u_{11}+u_{22}-2u_{33}; \end{aligned} \tag{15.42}$$

and five v branches with $\Omega=\Delta(12/5)^{1/2}$):

$$\begin{aligned} &E^2=\frac{12\Delta^2}{5}+\frac{c_F^2k^2}{105}\left[20-\frac{2(6^{1/2})}{\pi-\arctan 2(6^{1/2})}\right], v_{12}+v_{21}, v_{11}-v_{22};\\ &E^2=\frac{12\Delta^2}{5}+\frac{c_F^2k^2}{210}\left[85+\frac{2(6^{1/2})}{\pi-\arctan 2(6^{1/2})}\right], v_{13}+v_{31}, v_{23}+v_{32};\\ &E^2=\frac{12\Delta^2}{5}+\frac{c_F^2k^2}{105}\left[50+\frac{2(6^{1/2})}{\pi-\arctan 2(6^{1/2})}\right], v_{11}+v_{22}-2v_{33}. \end{aligned} \tag{15.43}$$

We have obtained all the Bose-spectrum branches except those with $\Omega=2\Delta$ by expanding the coefficients of the tensors A_{ij}, B_{ijab} at small k (15.19), (15.20). This procedure, however, cannot be used for the branches with $\Omega=2\Delta$, since the function $h(\omega)$ (and also $f(\omega)$, $g(\omega)$) has a singularity $\sim(\omega^2+4\Delta^2)^{-1/2}$ at $\omega^2\to-4\Delta^2$. Therefore, the branches with $\Omega=2\Omega$ call for a special investigation.

We start with the branch corresponding to the variable $u_{11}+u_{22}+u_{33}$. We separate from the quadratic form (15.16) the terms corresponding to the indicated variable by putting $u_{ia} = c(p)\delta_{ia}$, $v_{ia} = 0$. In this case

$$\delta_{ia}\delta_{ja}n_{1i}n_{1j} = \delta_{ia}\delta_{jb}(2n_{1a}n_{1b} - \delta_{ab})n_{1i}n_{1j} = 1$$

and (15.10) is transformed into

$$\sum_p c^2(p)A(p, u_{11} + u_{22} + u_{33}), \tag{15.44}$$

where

$$A(p, u_{11} + u_{22} + u_{33})$$
$$= \frac{4Z^2}{\beta V} \sum_{p_1 + p_2 = p} \left[\frac{(\xi_1 + i\omega_1)(\xi_2 + i\omega_2) - \Delta^2}{(\omega_1^2 + \xi_1^2 + \Delta^2)(\omega_2^2 + \xi_2^2 + \Delta^2)} - \frac{1}{\omega_1^2 + \xi_1^2 + \Delta^2} \right]. \tag{15.45}$$

Using the Feynman procedure to calculate $A(p, u_{11} + u_{22} + u_{33})$ and integrating with respect to ω_1 and ξ_1, we obtain

$$A(p, u_{11} + u_{22} + u_{33}) = \frac{Z^2 k_F^2}{4\pi^3 c_F} \int F \, d\Omega_1, \tag{15.46}$$

where

$$F = -\int_0^1 d\alpha [\ln(1 + \Delta^{-2}\alpha(1-\alpha)(\omega^2 + c_F^2(\mathbf{nk})^2)) + 2]. \tag{15.47}$$

The coefficient functions corresponding to the variables $v_{12} - v_{21}, v_{13} - v_{31}$ and $v_{23} - v_{32}$ are obtained in a similar manner.

They can be written in the form

$$A(p, v_{ik} - v_{ki}) = \frac{Z^2 k_F^2}{4\pi^3 c_F} \int (n_{1i}^2 + n_{1k}^2) F \, d\Omega_1. \tag{15.48}$$

We calculate the integral (15.47) for F:

$$F = -\frac{2a}{b} \arctan\frac{b}{a} = -\frac{a}{b}\left(\pi - 2\arctan\frac{a}{b}\right), \tag{15.49}$$

where

$$a^2 = \Delta^2 + \tfrac{1}{4}(\omega^2 + c_F^2(\mathbf{nk})^2),$$
$$b^2 = -\tfrac{1}{4}(\omega^2 + c_F^2(\mathbf{nk})^2). \tag{15.50}$$

As $\omega^2 \to -4\Delta^2$, the quantity b is positive and close to Δ^2; $a^2 = O(k^2)$, so that $2\arctan(a/b) \ll \pi$, and in the first-order approximation $F \approx -\pi a/\Delta$. This leads to an equation that determines in the first order the dispersion of the branch corresponding to the variable $u_{11} + u_{22} + u_{33}$:

$$\int a \, d\Omega_1 = 0. \tag{15.51}$$

We direct k along the third axis and denote $\cos\theta_1 = x$. Then $(\mathbf{n}_1\mathbf{k}) = kx$ and (15.45) becomes

$$\int_{-1}^{1} dx(4\Delta^2 - E^2 + c_F^2k^2x^2)^{1/2} = 0. \tag{15.52}$$

Putting

$$z^2 = \frac{4\Delta^2 - E^2}{c_F^2k^2} \tag{15.53}$$

we obtain the equation

$$\int_{-1}^{1} dx(x^2 + z^2)^{1/2} = 0 \tag{15.54}$$

or

$$(1+z^2)^{1/2} + \frac{z^2}{2}\ln\frac{(1+z^2)^{1/2}+1}{(1+z^2)^{1/2}-1} = 0. \tag{15.55}$$

Putting also

$$t = \ln\frac{(1+z^2)^{1/2}+1}{(1+z^2)^{1/2}-1}. \tag{15.56}$$

We can rewrite (15.55) in the simple form

$$\sinh t + t = 0. \tag{15.57}$$

If t is a nontrivial root of this equation, on substituting $z = (\sinh(t/2))^{-1}$ in (15.53) we find

$$E^2 = 4\Delta^2 - \frac{c_F^2k^2}{\sinh^2(t/2)}. \tag{15.58}$$

Equation (15.57) and the dispersion law (15.58) turn out to be the same as for the single nonphonon branch of the spectrum in the Fermi gas model with the s pairing investigated in section 13. It was indicated there that physical meaning can be attached to the branch (15.58), the first one to appear in the course of the analytic continuation with respect to the variable E from the upper half plane to the unphysical sheet. This branch is obtained if t is replaced by the smallest modulus nontrivial ($\neq 0$) root of (15.57), which is equal to

$$t_1 \approx 2.251 + i4.212. \tag{15.59}$$

The branch obtained was called 'resonant excitation'. It corresponds to the pole of the Green function of the Bose fields $c_{ia}(p)$ which is located near the branch point $E^2 = 4\Delta^2$.

We can treat in a similar way the remaining three branches with $\Omega = 2\Delta$. They correspond to the equations

$$\int_{-1}^{1} dx(x^2 + z^2)^{1/2}(1 \mp x^2) = 0 \tag{15.60}$$

or

$$\mp 2(1 + z^2)^{3/2} + (4 \mp z^2)\left[(1 + z^2)^{1/2} + \frac{z^2}{2}\ln\frac{(1 + z^2)^{1/2} + 1}{(1 + z^2)^{1/2} - 1}\right] = 0, \tag{15.61}$$

in which it is necessary to take the minus sign for the variable $v_{12} - v_{21}$ and the plus sign for $v_{13} - v_{31}, v_{23} - v_{32}$. Using (15.56), we obtain in place of (15.61)

$$\frac{\cosh t - 2}{2\cosh t - 1}\sinh t + t = 0, \quad v_{12} - v_{21}; \tag{15.62}$$

$$\frac{3\cosh t - 2}{2\cosh t - 3}\sinh t + t = 0, \quad v_{13} - v_{31}, v_{23} - v_{32}. \tag{15.63}$$

As a result, the dispersion laws for all branches with $\Omega = 2\Delta$ are given by formula (15.58), where t are nontrivial roots of (15.57) for the branch $u_{11} + u_{22} + u_{33}$, of (15.62) for $v_{12} - v_{21}$, of (15.63) for $v_{13} - v_{31}$, $v_{23} - v_{32}$. The branches with direct physical meaning are those appearing first in the analytic continuation from the physical sheet. For (15.62), the desired nontrivial solution with minimal modulus is of the form

$$t_2 \approx 2.93 + i4.22, \tag{15.64}$$

and for (15.63) is of the form

$$t_3 \approx 1.94 + i4.14. \tag{15.65}$$

The coefficients of $c_F^2 k^2$ are complex for all the branches with $\Omega = 2\Delta$ as noted above. The physical reason for this is the possibility of the decay of the Bose excitation into two fermions.

Let us list the results obtained above for the square of the Bose spectrum:

$u_{11} + u_{22} + u_{33}$	$4\Delta^2 + (0.237 - i0.295)c_F^2 k^2$
$u_{12} - u_{21}$	$c_F^2 k^2/5$
$u_{13} - u_{31}, u_{23} - u_{32}$	$2c_F^2 k^2/5$
$u_{11} - u_{22}, u_{12} + u_{21}$	$8\Delta^2/5 + 0.224c_F^2 k^2$
$u_{13} + u_{31}, u_{23} + u_{32}$	$8\Delta^2/5 + 0.388c_F^2 k^2$
$u_{11} + u_{22} - 2u_{33}$	$8\Delta^2/5 + 0.442c_F^2 k$

$$\begin{array}{ll}
v_{11}+v_{22}+v_{33} & c_F^2k^2/3 \\
v_{12}-v_{21} & 4\Delta^2+(0.111-i0.169)c_F^2k^2 \\
v_{13}-v_{31},\, v_{23}-v_{32} & 4\Delta^2+(0.353-i0.336)c_F^2k^2 \\
v_{11}-v_{22},\, v_{11}+v_{21} & 12\Delta^2/5+0.164c_F^2k^2 \\
v_{13}+v_{31},\, v_{23}+v_{32} & 12\Delta^2/5+0.418c_F^2k^2 \\
v_{11}+v_{22}-2v_{33} & 12\Delta^2/5+0.502c_F^2k^2.
\end{array} \tag{15.66}$$

Here we have replaced the coefficients of the type $(20+2(6^{1/2})\arctan 2(6^{1/2}))/105$ in front of $c_F^2k^2$ by their numerical values.

Let us briefly discuss the physical meaning of the modes obtained. The mode corresponding to the variable $v_{11}+v_{22}+v_{33}$ is the acoustic (sound) mode. The modes $u_{ik}-u_{ki}$ correspond to spin waves (longitudinal) $(u_{12}-u_{21})$ and transverse $(u_{13}-u_{31}, u_{23}-u_{32})$. The remaining 14 branches have a gap at $k=0$ and constitute different oscillation modes of the auxiliary field $c_{i\alpha}$. Five variables $u_{11}-u_{22}$, $u_{12}+u_{21}$, $u_{13}+u_{31}$, $u_{23}+u_{32}$, $u_{11}+u_{22}-2u_{33}$ are connected with the so-called 'real squashing mode' which is five-fold degenerate at $k=0$. Five other variables $v_{11}-v_{22}$, $v_{12}+v_{21}, v_{13}+v_{31}, v_{23}+v_{32}, v_{11}+v_{22}-2v_{33}$ give the squashing modes. It is also five-fold degenerate at $k=0$. Four modes with $\Omega=2\Delta$ are the so-called 'pair-breaking modes'. Their dispersion coefficients are complex, and this is connected with the possibility of decay of this collective mode into two fermions

Notice that most of these modes have been observed in different experiments.

To conclude this section we discuss the problem of stability of phonon branches of the Bose spectrum. We may consider stability of a given excitation under different processes, for instance under the decay into two fermions or the decay to two or more excitations of the same or different types.

In the isotropic *B*-phase it is impossible for a phonon to decay into two fermions because the phonon energy is much smaller than 2Δ. The decay of a given phonon excitation to two or more excitations of the same type is impossible kinematically provided that $d^2E/dk^2<0$ and the graph of $E(k)$ is located below the tangent (see Fig. 15.1).

This is equivalent to the positivity of the dispersion coefficient in the dispersion law

$$E(k)=uk(1-\gamma k^2). \tag{15.67}$$

This can easily be shown by using the energy–momentum conservation law:

$$E(k)=E(k_1)+E(|\mathbf{k}-\mathbf{k}_1|). \tag{15.68}$$

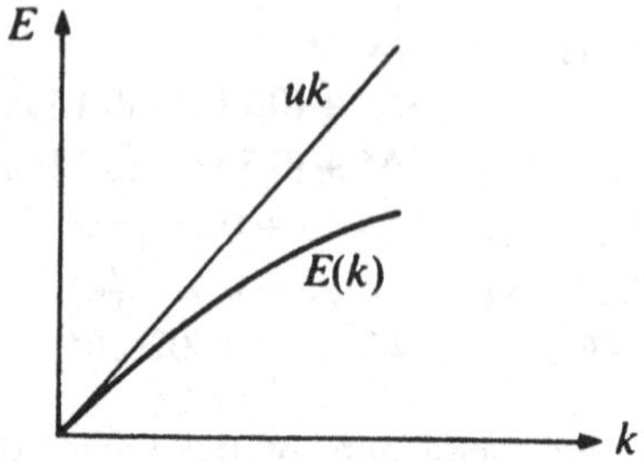

Fig. 15.1

According to (15.67) this is equivalent to the equation

$$k - \gamma k^3 = k_1 - \gamma k_1^3 + |\mathbf{k} - \mathbf{k}_1| - \gamma |\mathbf{k} - \mathbf{k}_1|^3 \tag{15.69}$$

or to

$$k - k_1 - |\mathbf{k} - \mathbf{k}_1| = \gamma(k^3 - k_1^3 - |\mathbf{k} - \mathbf{k}_1|^3). \tag{15.70}$$

For small k the angle θ between $\mathbf{k}$ and the decay phonon momentum $\mathbf{k}_1$ is small. So (for small θ) we have

$$|\mathbf{k} - \mathbf{k}_1|^2 = k^2 + k_1^2 - 2kk_1 \cos\theta = (k - k_1)^2 + 2kk_1(1 - \cos\theta)$$

$$k - k_1 - |\mathbf{k} - \mathbf{k}_1| = k - k_1 - [(k - k_1)^2 + 2kk_1(1 - \cos\theta)]^{1/2}$$

$$\approx -\frac{kk_1}{k - k_1}(1 - \cos\theta). \tag{15.71}$$

Let us substitute (15.71) into the left-hand side of (15.70). On the right-hand side we can put $|\mathbf{k} - \mathbf{k}_1| \approx k - k_1$. As a result we obtain

$$\gamma = -\frac{1 - \cos\theta}{3(k - k_1)^2} < 0. \tag{15.72}$$

So the decay is possible only for the case of a negative dispersion coefficient γ. We will show that all the phonon modes in the B-phase of ^{3}He are stable due to the positivity of dispersion coefficients γ for all these branches.

In order to obtain γ, we have to calculate the coefficient tensors $A_{ij}(p)$, $B_{ijab}(p)$ given in (15.19) and (15.20) at small $\omega, \mathbf{k}$ more accurately than we have done so far. Expanding the integrands in (15.19) and (15.20) up to q^4, where $q^2 = \omega^2 + c_F^2(\mathbf{n}_1\mathbf{k})^2$, we have

$$A_{ij}(p) - A_{ij}(0)$$

$$\approx \frac{Z^2 k_F^2}{4\pi^3 c_F} \int d\Omega_1 \left(\frac{q^2}{3\Delta^2} - \frac{q^4}{20\Delta^4} \right) n_{1i} n_{1j},$$

$$B_{ijab}(p) - B_{ijab}(0) \approx \frac{Z^2 k_F^2}{4\pi^3 c_F} \int d\Omega_1 \left(\frac{q^2}{6\Delta^2} - \frac{q^4}{30\Delta^4} \right) (\delta_{ab} - 2n_{1a}n_{1b}) n_{1i} n_{1j}. \tag{15.73}$$

Here the integrals $\int d\Omega_1$ can easily be calculated. We find the following formula for (15.5) for small p:

$$\begin{aligned} -\frac{Z^2 k_F^2}{4\pi^3 c_F} \sum_p \Bigg\{ & (\tfrac{1}{3} \pm \tfrac{1}{5}) w_{ia} w_{ia} \pm \tfrac{2}{15}(w_{aa} w_{bb} + w_{ia} w_{ai}) \\ & + w_{ia} w_{ja} \left[\frac{\delta_{ij}}{3} \left(\frac{\omega^2}{3\Delta^2} - \frac{\omega^4}{20\Delta^4} \right) + \frac{c_F^2}{15} (k^2 \delta_{ij} + k_i k_j) \left(\frac{1}{3\Delta^2} - \frac{\omega^2}{10\Delta^4} \right) \right. \\ & \left. - \frac{c_F^4}{700\Delta^4} (k^4 \delta_{ij} + 4k^2 k_1 k_j) \right] \pm w_{ia} w_{jb} \left[\delta_{ab} \left[\frac{\delta_{ij}}{3} \left(\frac{\omega^2}{6\Delta^2} - \frac{\omega^4}{30\Delta^4} \right) \right. \right. \\ & \left. + \frac{c_F^2}{15} (k^2 \delta_{ij} + 2k_i k_j) \left(\frac{1}{6\Delta^2} - \frac{\omega^2}{15\Delta^4} \right) - \frac{c_F^4}{1050} (k^4 \delta_{ij} + 4k^2 k_i k_j) \right] \\ & + (\delta_{ab}\delta_{ij} + \delta_{ai}\delta_{bj} + \delta_{aj}\delta_{bi}) \left[\frac{2}{15} \left(-\frac{\omega^2}{6\Delta^2} + \frac{\omega^4}{30\Delta^4} \right) + \frac{2k^2}{105} \left(-\frac{c_F^2}{6\Delta^2} + \frac{\omega^2 c_F^2}{15\Delta^4} \right) \right. \\ & \left. + \frac{c_F^4 k^4}{4725\Delta^4} \right] + \tfrac{4}{315} (\delta_{ij} k_a k_b + \delta_{ab} k_i k_j + \delta_{ai} k_b k_j \\ & \left. + \delta_{bj} k_a k_i + \delta_{aj} k_b k_i + \delta_{bi} k_a k_j) \left(-\frac{c_F^2}{2\Delta^2} + \frac{\omega^2 c_F^2}{5\Delta^4} \right) + \frac{8c_F^4}{4725\Delta^4} k_a k_b k_i k_j \right] \Bigg\}. \end{aligned} \tag{15.74}$$

Here we have to substitute $w_{ia} = u_{ia}$ and take the upper sign of $\pm, \mp$, and then put $w_{ia} = v_{ia}$ and take the lower sign. The quadratic form (15.74) determines the phonon branches of collective excitations in the B-phase; this phase is isotropic. So we may only consider excitations expanding in a given direction, say along the third axis. If we put $k_1 = k_2 = 0$, $k_3 = k$, the quadratic form of the variables w_{ia} breaks up into a sum of four independent forms. The first of them depends on w_{12}, w_{21}, the second on w_{13}, w_{31}, the third on w_{23}, w_{32}, and the fourth one depends on w_{11}, w_{22}, w_{33}.

If $w_{ia} = v_{ia}$, the phonon branch is defined by a form depending on v_{11}, v_{22}, v_{33}. This form is proportional to

$$a(v_{11}^2 + v_{22}^2) + bv_{33}^2 + 2cv_{11}v_{22} + 2d(v_{11} + v_{22})v_{33}. \tag{15.75}$$

Here

$$\begin{aligned}
&a = \tfrac{4}{15} + a_1, \quad b = \tfrac{4}{15} + b_1, \quad c = -\tfrac{2}{15} + c_1, \quad d = -\tfrac{2}{15} + d_1, \\
&a_1 = \tfrac{11}{90}x^2 + \tfrac{13}{630}y^2 - \tfrac{17}{900}x^4 - \tfrac{19}{3150}x^2y^2 - \tfrac{1}{900}y^4, \\
&b_1 = \tfrac{11}{90}x^2 + \tfrac{17}{210}y^2 - \tfrac{17}{900}x^4 - \tfrac{9}{350}x^2y^2 - \tfrac{37}{3780}y^4, \\
&c_1 = \tfrac{1}{45}x^2 + \tfrac{1}{315}y^2 - \tfrac{1}{225}x^4 - \tfrac{2}{1575}x^2y^2 - \tfrac{1}{4775}y^4, \\
&d_1 = \tfrac{1}{45}x^2 + \tfrac{1}{105}y^2 - \tfrac{1}{225}x^4 - \tfrac{2}{525}x^2y^2 - \tfrac{1}{945}y^4,
\end{aligned} \tag{15.76}$$

where $x = \omega/\Delta$, $y = c_F k/\Delta$. The third-order determinant of the quadratic from (15.75) is equal to $(a-c)\,(b(a+c) - 2d^2)$. We see that $a - c \neq 0$ for ω, $k \to 0$. So we obtain the equation

$$b(a+c) - 2d^2 = 0, \tag{15.77}$$

or

$$\tfrac{2}{15}(b_1 + 2a_1 + 2c_1 + 4d_1) + b_1(a_1 + c_1) - 2d_1^2 = 0. \tag{15.78}$$

We have

$$b_1 + 2a_1 + 2c_1 + 2d_1 = \frac{x^2}{2} + \frac{y^2}{6} - \frac{x^4}{12} - \frac{x^2y^2}{18} - \frac{y^4}{60},$$

$$b_1(a_1 + c_1) - 2d_1^2 = \frac{x^4}{60} + \frac{13x^2y^2}{945} + \frac{11y^4}{6300}. \tag{15.79}$$

Inserting (15.79) into (15.78) we get the equation

$$x^2 + \frac{y^2}{3} + \frac{x^4}{12} + \frac{2x^2y^2}{21} + \frac{y^4}{140} = 0. \tag{15.80}$$

The substitution $x = \omega/\Delta$, $y = c_F k/\Delta$, $i\omega \to E$, gives the sound branch of the spectrum in the form

$$E = \frac{c_F k}{3^{1/2}}\left(1 - \frac{2c_F^2 k^2}{45\Delta^2}\right). \tag{15.81}$$

This mode is stable under decays into two or more excitations of the same type. Let us note that the value of dispersion coefficient $\gamma = 2c_F^2/45\Delta^2$ is twice that of the dispersion coefficient for the superfluid Fermi gas with s pairing (see section 13).

Now let us consider forms of the variables (u_{12}, u_{21}), (u_{13}, u_{31}) which are proportional to the following expressions:

$$\tfrac{2}{15}(u_{12} + u_{21})^2 + (u_{12}^2 + u_{21}^2)\left(\frac{13x^2}{90} + \frac{19y^2}{630} - \frac{7x^4}{300} - \frac{31x^2y^2}{3150} - \frac{41y^4}{18900}\right)$$

$$-2u_{12}u_{21}\left(\frac{x^2}{45}+\frac{y^2}{315}-\frac{x^4}{225}-\frac{2x^2y^2}{1575}-\frac{y^4}{4725}\right)$$

$$\tfrac{2}{15}(u_{13}+u_{31})^2+u_{13}^2\left(\frac{13x^2}{90}+\frac{y^2}{42}-\frac{7x^4}{300}-\frac{23x^2y^2}{3150}-\frac{y^4}{756}\right)$$

$$+u_{31}^2\left(\frac{13x^2}{90}+\frac{19y^2}{210}-\frac{7x^4}{300}-\frac{31x^2y^2}{1050}-\frac{41y^4}{3780}\right)$$

$$-2u_{13}u_{31}\left(\frac{x^2}{45}+\frac{y^2}{105}-\frac{x^4}{225}-\frac{2x^2y^2}{525}-\frac{y^4}{945}\right). \qquad (15.82)$$

The form of the variables (u_{23}, u_{32}) can be obtained from the second form in (15.82) by replacing (u_{13}, u_{31}) by (u_{23}, u_{32}). Equating the determinants of the forms in (15.82) to zero we obtain one branch of longitudinal spin waves (variables u_{12}, u_{21}):

$$E=\frac{c_F k}{5^{1/2}}\left(1-\frac{2c_F^2k^2}{105\Delta^2}\right), \qquad (15.83)$$

and a two-fold degenerate branch of transverse spin waves (variables u_{13}, u_{31} and u_{23}, u_{32}):

$$E=\left(\frac{2}{5}\right)^{1/2} c_F k\left(1-\frac{173c_F^2k^2}{3360\Delta^2}\right). \qquad (15.84)$$

The branches (15.83) and (15.84) turn out to be stable, as does the sound branch (15.81).

16

Collective excitations in the A-phase of ^{3}He

The A-phase is anisotropic, and there exists a preferred direction along the total orbital momentum of Cooper pairs. The gap in the Fermi spectrum has the form

$$\Delta = \Delta_0 \sin\theta, \quad \Delta_0 = 2cZ. \tag{16.1}$$

The constant c enters in the expression for the condensate wave function $c_{ia}^{(0)}$:

$$c_{ia}^{(0)}(p) = c(\beta V)^{1/2}\delta_{p0}(\delta_{i1} + i\delta_{i2})\delta_{a3}. \tag{16.2}$$

One can find c from the gap equation

$$g_0^{-1} + \frac{2Z^2}{\beta V}\sum_p \sin^2\theta(\omega^2 + \xi^2 + 4c^2Z^2\sin^2\theta)^{-1} = 0. \tag{16.3}$$

Performing the shift $c_{ia}(p) \to c_{ia}(p) + c_{ia}^{(0)}(p)$, we can extract a quadratic form from

$$Q = g_0^{-1}\sum_p c_{ia}^*(p)c_{ia}(p) - \tfrac{1}{4}\operatorname{Tr}(Gu)^2, \tag{16.4}$$

where

$$u_{p_1p_2} = (\beta V)^{-1/2}\begin{pmatrix} 0, (n_{1i} - n_{2i})\sigma_a c_{ia}(p_1 + p_2) \\ -(n_{1i} - n_{2i})\sigma_a c_{ia}^*(p_1 + p_2), 0 \end{pmatrix} \tag{16.5}$$

$$G_{p_1p_2}^{-1} = \begin{pmatrix} Z^{-1}(i\omega - \xi)\delta_{p_1p_2}, 2c\sigma_3(n_1 + in_2)\delta_{p_1+p_2,0} \\ -2\bar{c}\sigma_3(n_1 - in_2)\delta_{p_1+p_2,0}, Z^{-1}(\xi - i\omega)\delta_{p_1p_2} \end{pmatrix}. \tag{16.6}$$

Inverting G^{-1} we obtain

$$G = \frac{Z}{M}\begin{pmatrix} -(i\omega + \xi)\delta_{p_1p_2}, \Delta_0\sigma_3(n_1 + in_2)\delta_{p_1+p_2,0} \\ -\Delta_0\sigma_3(n_1 - in_2)\delta_{p_1+p_2,0}, (i\omega - \xi)\delta_{p_1p_2} \end{pmatrix}, \tag{16.7}$$

$$M = \omega^2 + \xi^2 + \Delta^2 = \omega^2 + \xi^2 + \Delta_0^2\sin^2\theta.$$

Then

$$-\tfrac{1}{4}\operatorname{Tr}(Gu)^2$$

$$= -\frac{1}{4}\sum_{p_1,p_2,p_3,p_4} \operatorname{tr}(G_{p_1p_2}u_{p_2p_3}G_{p_3p_4}u_{p_4p_1})$$

$$= -\frac{Z^2}{4\beta V}\sum_p \Big\{ c^*_{ia}(p)c^*_{jb}(-p)\Big[4Z^2c^2 \sum_{p_3-p_1=p} \operatorname{tr}(\sigma_3\sigma_a\sigma_3\sigma_b)$$
$$\times M_1^{-1}M_3^{-1}(n_1+in_2)_1(-n_1-n_3)_i(n_1+in_2)_3(-n_3-n_1)_j\Big]$$
$$+ c_{ia}(p)c_{jb}(-p)\Big[4Z^2\bar{c}^2 \sum_{p_3-p_1=p} \operatorname{tr}(\sigma_3\sigma_a\sigma_3\sigma_b)M_1^{-1}M_3^{-1}$$
$$\times (n_1-in_2)_1(-n_1-n_3)_i(n_1-in_2)_3(-n_3-n_1)_j\Big] + 2c^*_{ia}(p)c_{jb}(p)$$
$$\times \sum_{p_1+p_2=p} [\operatorname{tr}(\sigma_a\sigma_b)(-n_1+n_3)_i(n_1-n_3)_j(\xi_1+i\omega_1)$$
$$\times(\xi_3+i\omega_3)M_1^{-1}M_3^{-1}]\Big\}.$$

We shall be mainly interested in the region of small p. In this region we may replace $\mathbf{n}_3$ by $-\mathbf{n}_1$ if $p_1+p_3=p$, or $\mathbf{n}_3$ by $\mathbf{n}_1$ if $p_3-p_1=p$. Replacing p_1 by $-p_1$ in the sums with $p_3-p_1=p$ and taking the trace we get

$$\sum_p\Big\{c^*_{ia}(p)c_{ja}(p)\frac{4Z^2}{\beta V}\sum_{p_1+p_2=p} G_1G_2(\xi_1+i\omega_1)(\xi_2+i\omega_2)n_{1i}n_{1j}$$
$$+[c^*_{ia}(p)c^*_{jb}(-p)c^2+c_{ia}(p)c_{jb}(-p)\bar{c}^2]\frac{2Z^2}{\beta V}\sum_{p_1+p_2=p}(n_1\pm in_2)_1^2$$
$$\times n_{1i}n_{1j}G_1G_2(2\delta_{a3}\delta_{b3}-\delta_{ab})\Big\}.$$

We may take c real. Adding $g_0^{-1}\sum_p c^*_{ia}(p)c_{ia}(p)$, we obtain the quadratic form Q of S_{eff},

$$\sum_p c^*_{ia}(p)c_{ja}(p)\Big[\delta_{ij}g_0^{-1}+\frac{4Z^2}{\beta V}\sum_{p_1+p_2=p}G(p_1)G(p_2)(\xi_1+i\omega_1)$$
$$\times(\xi_2+i\omega_2)n_{1i}n_{ij}\Big]+\sum_p\Big(\sum_{a=1,2}c^*_{ia}(p)c^*_{jb}(-p)$$
$$+c_{ia}(p)c_{jb}(-p)\Big)-(c^*_{i3}(p)c^*_{j3}(-p)+c_{i3}(p)c_{j3}(-p))$$
$$\times\frac{2Z^2\Delta_0^2}{\beta V}\sum_{p_1+p_2=p}(n\pm in_2)^2n_{1i}n_{1j}G(p_1)G(p_2), \tag{16.8}$$

where

$$G(p)=(\omega^2+\xi^2+\Delta_0^2\sin^2\theta)^{-1}. \tag{16.9}$$

The alternative sign $\pm$ in (16.8) means that the + is chosen when the sum is multiplied by $c^*_{ia}c^*_{jb}$ and that − is chosen when it is multiplied by $c_{ia}c_{jb}$.

Expression (16.8) is the sum of three independent forms, which differ from one another by the value of the spin index a, and go into one another after a corresponding change of variables. This is why the spectrum of the A-phase in this model is three-fold degenerate, and we may consider one of these three forms, say the form with $a=1$.

So let us take the form with $a=1$ and consider the terms with $p=0$:

$$-\frac{Z^2k_F^2}{2\pi^2c_F}[3v_{11}^2+3u_{21}^2-2u_{21}v_{11}+(u_{11}-v_{21})^2]. \qquad (16.10)$$

This expression does not contain the variables u_{31}, $u_{11}+v_{21}$. Hence the variables

$$u=u_{31}, \quad v=v_{31}, \quad w=\tfrac{1}{2}(u_{11}+v_{21}) \qquad (16.11)$$

can be chosen as 'phonon' variables. Associated with them are branches of the Bose spectrum that are phonon-like, i.e. $\lim_{k\to 0} E(k)=0$. The phonon Bose spectrum is defined by the form of the phonon variables. In this form we can replace $A_{ij}(p)$ by $A_{ij}(p)-A_{ij}(0)$, $B_{ij}(p)$ by $B_{ij}(p)-B_{ij}(0)$, due to $A_{ij}(0)=B_{ij}(0)=0$ for the phonon variables.

If we use the Feynman trick and integrate with respect to ω_1, ξ_1 $(T\to 0)$, we obtain

$$A_{ij}(p)-A_{ij}(0)=\frac{Z^2k_F^2}{4\pi^3c_F}\int_0^1 d\alpha\int d\Omega_1 n_{1i}n_{1j}$$
$$\times\left[\ln\left(1+\frac{\alpha(1-\alpha)q^2}{\Delta_0^2\sin^2\theta}\right)+\frac{\alpha(1-\alpha)q^2}{\Delta_0^2\sin^2\theta+\alpha(1-\alpha)q^2}\right], \qquad (16.12)$$
$$B_{ij}^{\pm}(p)-B_{ij}^{\pm}(0)=\frac{Z^2k_F^2}{4\pi^3c_F}\int_0^1 d\alpha\int d\Omega_1 n_{1i}n_{1j}e^{\pm 2i\omega_1}\frac{\alpha(1-\alpha)q^2}{\Delta_0^2\sin^2\theta+\alpha(1-\alpha)q^2}.$$

Here $q^2=\omega^2+c_F^2(\mathbf{nk})^2$, B_{ij}^+ is the coefficient of $c_i^*c_j^*$, B_{ij}^- is the coefficient of c_ic_j.

Let us expand the integrals in (16.12) in powers of $\alpha(1-\alpha)q^2/\Delta_0^2\sin^2\theta$. If we take only the first term of this expansion, we come to a logarithmically divergent integral when calculating $A_{33}(p)-A_{33}(0)$. This expression is proportional to

$$\int d\Omega_1\frac{q^2n_3^2}{\Delta_0^2\sin^2\theta_1}=\frac{2\pi(\omega^2+c_F^2k_3^2)}{\Delta_0^2}\int_0^\pi\frac{d\theta_1}{\sin\theta_1}+\text{finite part}. \qquad (16.13)$$

The origin of this divergence is that our expansion parameter $q^2\alpha(1-\alpha)/\Delta_0^2\sin^2\theta$ is not small near the poles of the Fermi sphere $(\theta_1 = 0, \pi)$. A more accurate calculation is needed here, and calculation gives the following result:

$$A_{33}(p) - A_{33}(0) = \frac{Z^2 k_F^2}{48\pi^2 c_F \Delta_0^2}\left[2p^2 \ln\frac{4\Delta_0^2}{p^2} + \frac{1}{3}p^3 + \frac{2}{3}c_F^2 k^2 - 2c_F^2 k_3^2\right] \tag{16.14}$$

for small $p^2 = \omega^2 + c_F^2 k_3^2$.

The remaining elements of $A_{ij}(p) - A_{ij}(0)$ and $B_{ij}(p) - B_{ij}(0)$ contain no divergences.

We thus find the following matrix of phonon variables (see (16.11)):

$$Q = \frac{Z^2 k_F^2}{48\pi^2 c_F \Delta_0^2}\begin{pmatrix} a(p) + \frac{c_F^2}{12}(k_1^2 - k_2^2), \frac{c_F^2}{6}k_1 k_2, c_F^2 k_1 k_3 \\ \frac{c_F^2}{6}k_1 k_2, a(p) + \frac{c_F^2}{12}(k_2^2 - k_1^2), c_F^2 k_2 k_3 \\ c_F^2 k_1 k_3, c_F^2 k_2 k_3, 3\omega^2 + c_F^2 k^2 \end{pmatrix} \tag{16.15}$$

where

$$a(p) = p^2\left(\ln\frac{4\Delta_0^2}{p^2} + \frac{1}{6}\right) + \frac{c_F^2}{3}(k^2 - 3k_3^2). \tag{16.16}$$

The equation $\det Q = 0$ may be written in the form

$$\left(a(p) - \frac{c_F^2 k_\perp^2}{12}\right)\left[(3\omega^2 + c_F^2 k^2)\left(a(p) + \frac{c_F^2 k_\perp^2}{12}\right) - c_F^4 k_\parallel^2 k_\perp^2\right] = 0, \tag{16.17}$$

where $k_\parallel^2 = k_3^2$, $k_\perp^2 = k_1^2 + k_2^2$. From (16.17) we obtain three branches of the spectrum: one $E^2 \approx c_F^2 k^2/3$ and two $E^2 \approx c_F^2 k_\parallel^2$. We can also find corrections to the linear dispersion law, as well as the domains of stability of the Bose spectrum. The result can be stated as follows:

$$E_1(k) = \frac{c_F k}{3^{1/2}}\left(1 - \frac{\sin^2\theta\cos^2\theta}{2(\cos^2\theta - 1/3)\ln 4\Delta_0^2/f_1(\theta, k)}\right),$$

$$E_2(k) = c_F k_\parallel\left(1 - \frac{11\cos^2\theta - 3}{24\cos^2\theta \ln 4\Delta_0^2/f_2(\theta, k)}\right),$$

$$E_3(k) = c_F k_\parallel\left(1 - \frac{51\cos^4\theta - 40\cos^2\theta + 5}{72\cos^2\theta(\cos^2\theta - 1/3)\ln 4\Delta_0^2/f_3(\theta, k)}\right), \tag{16.18}$$

where

$$f_1(\theta, k) = c_F^2 k^2(\cos^2\theta - 1|3),$$

$$f_2(\theta, k) = \frac{c_F^2 k^2}{12}(11\cos\theta - 3),$$

$$f_3(\theta, k) = \frac{c_F^2 k^2}{36}\frac{51\cos^4\theta - 40\cos^2\theta + 5}{\cos^2\theta(\cos^2\theta - 1/3)}. \tag{16.19}$$

According to these formulae, stability of the spectrum in the A-phase depends on the angle θ between the excitation momentum and the direction. The first (sound) branch is stable in the cones $\cos^2\theta > 1/3$; the second branch is stable in the cones $\cos^2\theta > 3/11$. The third branch is stable in the domains given by

$$\cos^2\theta > \frac{20 + 145^{1/2}}{51}, \quad \tfrac{1}{3} > \cos^2\theta > \frac{20 - 145^{1/2}}{51}.$$

Outside the stability domains the excitation energies become complex due to the imaginary parts of the logarithms in (16.18). Instability is connected with the possibility of decay of a given excitation into two fermions with momenta close to $\pm k_F \mathbf{e}_3$.

Now we turn to the nonphonon branches of the Bose spectrum in the A-phase at $T = 0$. The main result is that the energies $E(\mathbf{k})$ of all nonphonon excitations turn out to be complex with nonvanishing imaginary parts even at $\mathbf{k} = 0$. This result can easily be explained by the possibility of decay of a given nonphonon excitation into two fermions with momenta close to $\pm k_F \mathbf{e}_3$. We can obtain equations for all values of $E_\alpha(0)$ (complex energies on nonphonon branches) from the quadratic part Q of S_{eff} as in (16.8). This is a sum of three independent forms depending on c_{i1}, c_{i2}, c_{i3}. The second and third forms turn into the first one after the replacement $c_{i2} \to c_{i1}$, $c_{i3} \to ic_{i1}$. So it is sufficient to consider only the first form depending on c_{i1}, c_{i1}^*, which is equal to

$$\sum_p c_{i1}^*(p)c_{j1}(p)\left[\delta_{ij}g_0^{-1} + \frac{4Z^2}{\beta V}\sum_{p_1+p_2=p} n_{1i}n_{1j}G_1G_2(\xi_1 + i\omega_1)(\xi_2 + i\omega_2)\right]$$
$$+ (c_{i1}^*(p)c_{j1}^*(-p) + c_{i1}(p)c_{j1}(-p))\frac{2Z^2\Delta_0^2}{\beta V}\sum_{p_1+p_2=p}(n_1 \pm in_2)^2 n_{1i}n_{1j}G_1G_2, \tag{16.20}$$

where $G(p)$ is given by (16.9), $\Delta^2 = \Delta_0^2\sin^2\theta = \Delta_0^2(n_1^2 + n_2^2)$. We have to take the upper sign in $(n_1 \pm in_2)^2$ in (16.20) when multiplying by $c_{i1}^*c_{j1}^*$ and to take the lower sign when multiplying by $c_{i1}c_{j1}$.

We wish to obtain all the branches of the Bose spectrum which are defined by (16.20) at $k = 0$. In this case the form of variables c_{i1}, c_{j1}^* is the sum of the form depending on $c_{11}, c_{11}^*, c_{21}, c_{21}^*$ and the form depending on c_{31}, c_{31}^*. The coefficient functions in front of $c_{i1}^* c_{j1}$, $c_{i1}^* c_{j1}^*$, $c_{i1} c_{j1}$ $(i,j = 1,2)$ may be written down as follows:

$$\delta_{ij} g_0^{-1} + \frac{4Z^2}{\beta V} \sum_{p_1+p_2=p} n_{1i} n_{ij} (\xi_1 + i\omega_1)(\xi_2 + i\omega_2) G_1 G_2$$

$$= \frac{2\delta_{ij} Z^2}{\beta V} \sum_{p_1+p_2=p} (n_1^2 + n_2^2)[(\xi_1 + i\omega_1)(\xi_2 + i\omega_2) G_1 G_2 - G_1],$$

$$\frac{2Z^2\Delta_0^2}{\beta V} \sum_{p_1+p_2=p} (n_1 \pm i n_2)^2 n_{1i} n_{1j} G_1 G_2 = b_{ij} \frac{Z^2 \Delta_0^2}{2\beta V} \sum_{p_1+p_2=p} (n_1^2 + n_2^2) G_1 G_2. \tag{16.21}$$

Here b_{ij} $(i,j = 1,2)$ are the entries of the matrix

$$\begin{pmatrix} 1, & \pm i \\ \pm i, & -1 \end{pmatrix}, \tag{16.22}$$

in which the minus sign corresponds to the variables $c_{i1} c_{j1}$ and the plus sign to $c_{i1}^* c_{j1}^*$. Here we have used (16.3). Let $f(\omega)$ be the coefficient of δ_{ij} and $g(\omega)$ be the coefficient of b_{ij}.

We can write the quadratic form of the variables $u_{11}, u_{21}, v_{11}, v_{21}$ as a sum of two forms

$$(f(\omega) + g(\omega))(u_{11}^2 + v_{21}^2) - 2g(\omega) u_{11} v_{21}$$
$$+ (f(\omega) - g(\omega))(v_{11}^2 + u_{21}^2) - 2g(\omega) v_{11} u_{21}. \tag{16.23}$$

The corresponding matrices are

$$\begin{pmatrix} f(\omega) + g(\omega), & -g(\omega) \\ -g(\omega), & f(\omega) + g(\omega) \end{pmatrix}, \begin{pmatrix} f(\omega) - g(\omega), & -g(\omega) \\ -g(\omega), & f(\omega) - g(\omega) \end{pmatrix}. \tag{16.24}$$

Equating the determinants of these matrices to zero, we obtain the equations

$$f(f + 2g) = 0, \quad f(f - 2g) = 0,$$

or

$$f(\omega) = 0, \quad f(\omega) + 2g(\omega) = 0, \quad f(\omega) - 2g(\omega) = 0. \tag{16.25}$$

We also have to consider the equation corresponding to the form of the variables c_{31}, c_{31}^*. This equation can be written as follows:

$$h(\omega) = g_0^{-1} + \frac{4Z^2}{\beta V} \sum_{p_1+p_2=p} n_3^2(\xi_1 + i\omega_1)(\xi_2 + i\omega_2)G_1G_2$$
$$= \frac{2Z^2}{\beta V} \sum_{p_1+p_2=p} [2n_3^2(\xi_1 + i\omega_1)(\xi_2 + i\omega_2)G_1G_2 - (n_1^2 + n_2^2)G_1]. \qquad (16.26)$$

We may rewrite the three equations (16.25) in the form

$$\frac{2Z^2}{\beta V} \sum_{p_1+p_2=p} (n_1^2 + n_2^2)\{G_1G_2[(\xi_1 + i\omega_1)(\xi_2 + i\omega_2) \pm (1,0)\Delta^2] - G_1\} = 0, \qquad (16.27)$$

where $\pm(1,0)\Delta^2$ means Δ^2 or $-\Delta^2$ or 0. We can replace sums by integrals and substitute G_1, G_2 by their expressions. So we rewrite (16.26) and (16.27) as follows:

$$\frac{2Z^2k_F^2}{(2\pi)^4 c_F} \int d\Omega_1\, d\omega_1\, d\xi_1$$
$$\left[\frac{2\cos^2\theta_1(\xi_1 + i\omega_1)(\xi_2 + i\omega_2)}{(\omega_1^2 + \xi_1^2 + \Delta^2)(\omega_2^2 + \xi_2^2 + \Delta^2)} - \frac{\sin^2\theta_1}{\omega_1^2 + \xi_1^2 + \Delta^2}\right] = 0$$
$$\frac{2Z^2k_F^2}{(2\pi)^4 c_F} \int \sin^2\theta_1\, d\Omega_1\, d\omega_1 d\xi_1$$
$$\left[\frac{(\xi_1 + i\omega_1)(\xi_2 + i\omega_2) \pm (1,0)\Delta^2}{(\omega_1^2 + \xi_1^2 + \Delta^2)(\omega_2^2 + \xi_2^2 + \Delta^2)} - \frac{1}{\omega_1^2 + \xi_1^2 + \Delta^2}\right] = 0. \qquad (16.28)$$

Using the Feynman trick and integrating with respect to ω_1, ξ_1, we obtain

$$\frac{Z^2k_F^2}{4\pi^3 c_F} \int_0^1 d\alpha \int \cos^2\theta d\Omega \left[\ln(1 + \Delta^{-2}\alpha(1-\alpha)\omega^2) + \frac{\alpha(1-\alpha)\omega^2}{\Delta^2 + \alpha(1-\alpha)\omega^2}\right] = 0,$$
$$\frac{Z^2k_F^2}{4\pi^3 c_F} \int_0^1 d\alpha \int \sin^2\theta d\Omega \left[\ln(1 + \Delta^{-2}\alpha(1-\alpha)\omega^2)\right.$$
$$\left. + \frac{2\alpha(1-\alpha)\omega^2 + \Delta^2 \mp (1,0)\Delta^2}{\Delta^2 + \alpha(1-\alpha)\omega^2}\right] = 0 \qquad (16.29)$$

instead of (16.28). Now we can integrate with respect to α:

$$\int_0^1 d\alpha \left[\ln(1 + \Delta^{-2}\alpha(1-\alpha)\omega^2) + \frac{\alpha(1-\alpha)\omega^2}{\Delta^2 + \alpha(1-\alpha)\omega^2}\right]$$
$$= \frac{\omega^2 + 2\Delta^2}{\omega(\omega^2 + 4\Delta^2)^{1/2}} \ln\frac{(\omega^2 + 4\Delta^2)^{1/2} + \omega}{(\omega^2 + 4\Delta^2)^{1/2} - \omega} - 1$$

$$\int_0^1 \frac{d\alpha\, \alpha(1-\alpha)}{\Delta^2+\alpha(1-\alpha)\omega^2} = 1 - \frac{2\Delta^2}{\omega(\omega^2+4\Delta^2)^{1/2}} \ln \frac{(\omega^2+4\Delta^2)^{1/2}+\omega}{(\omega^2+4\Delta^2)^{1/2}-\omega}. \qquad (16.30)$$

Let us substitute (16.30) into (16.29), replace ω by $\Delta_0\omega$ and let $\cos\theta = x$. We obtain the equations

$$\int_0^1 dx(1-x^2)\frac{\omega^2+4(1+x^2)}{\omega(\omega^2+4(1-x^2))^{1/2}} \ln \frac{(\omega^2+4(1-x^2))^{1/2}+\omega}{(\omega^2+4(1-x^2))^{1/2}-\omega} = 0,$$

$$\int_0^1 dx(1-x^2)\frac{\omega^2+2(1-x^2)}{\omega(\omega^2+4(1-x^2))^{1/2}} \ln \frac{(\omega^2+4(1-x^2))^{1/2}+\omega}{(\omega^2+4(1-x^2))^{1/2}-\omega} = 0,$$

$$\int_0^1 dx(1-x^2)\frac{\omega}{(\omega^2+4(1-x))^{1/2}} \ln \frac{(\omega^2+4(1-x^2))^{1/2}+\omega}{(\omega^2+4(1-x^2))^{1/2}-\omega} = 0,$$

$$\int_0^1 dx\, x^2\left[\frac{\omega^2+2(1-x^2)}{\omega(\omega^2+4(1-x^2))^{1/2}} \ln \frac{(\omega^2+4(1-x^2))^{1/2}+\omega}{(\omega^2+4(1-x^2))^{1/2}-\omega} - 1\right] = 0.$$

The first of these equations is $f(\omega)-2g(\omega)=0$, the second is $f(\omega)=0$, the third $f(\omega)+2g(\omega)=0$ and the fourth $h(\omega)=0$. These equations give the Bose spectrum after the analytic continuation $i\omega \rightarrow E/\Delta_0$. The branches corresponding to the second and the fourth equations are two-fold degenerate.

The third and the fourth equations in (16.31) have solutions corresponding to the phonon branches. The first and second equations lead to the complex energies of nonphonon modes. Computer calculations give

$$E_1(0) = \Delta_0(1.96 - i0.31), \quad E_2(0) = \Delta_0(1.17 - i0.13). \qquad (16.32)$$

The second branch $E_2(0)$ is two-fold degenerate.

Now let us write down all the branches of the Bose spectrum of the model in the A-phase with their corresponding variables:

$$E = \frac{c_F}{3^{1/2}}k, \quad u_{11}+v_{21},\ u_{12}+v_{22},\ u_{13}-v_{23};$$

$$E = c_F k_\parallel, \quad u_{31},\ v_{31},\ u_{32},\ v_{32},\ u_{33},\ v_{33};$$

$$E = \Delta_0(1.96 - i0.31), \quad u_{21}+v_{11},\ u_{22}+v_{12},\ u_{23}-v_{13};$$

$$E = \Delta_0(1.17 - i0.13), \quad u_{11}-v_{21},\ u_{21}-v_{11},\ u_{12}-v_{22},\ u_{22}-v_{12},$$

$$u_{13}-v_{23},\ u_{23}+v_{13}. \qquad (16.33)$$

Here we neglect the corrections to the energies of phonon modes $c_F k/3^{1/2}$, $c_F k_\parallel$ given by (16.18).

17

Superfluidity and Bose excitations in ^{3}He films

In this section we consider a model of ^{3}He film – a two-dimensional analogue of the three-dimensional ^{3}He model discussed in the previous sections. Several superfluid phases turn out to be possible in the model. Two of them, denoted below as a and b are energetically advantageous and stable relative to small fluctuations. The Bose spectrum of the model may be analysed by the method developed above for the three-dimensional model. The spectrum contains both the phonon (Goldstone) branches, whose number varies for different phases, and the nonphonon branches which have an energy gap.

The ^{3}He film model is described by the effective action functional

$$S_{\text{eff}} = g_0^{-1} \sum_{p,i,a} c_{ia}^*(p)c_{ia}(p) + \tfrac{1}{2}\ln\det M(c,c^*)/M(0,0), \tag{17.1}$$

where M is an operator of the form

$$M = \begin{pmatrix} Z^{-1}(i\omega - \xi + \mu(\boldsymbol{\sigma},\mathbf{H}))\delta_{p_1p_2}, & (\beta V)^{-1/2}(n_{1i} - n_{2i})\sigma_a c_{ia}(p_1 + p_2) \\ -(\beta V)^{-1/2}(n_{1i} - n_{2i})\sigma_a c_{ia}^*(p_1 + p_2), & Z^{-1}(-i\omega + \xi + \mu(\boldsymbol{\sigma},\mathbf{H}))\delta_{p_1p_2} \end{pmatrix}. \tag{17.2}$$

In the two-dimensional model the index i may take only two values $i = 1, 2$. This is the main difference between the two-dimensional model and the model of three-dimensional ^{3}He. In all other respects the notation used here is identical with that used in the previous sections. In particular, $c_{ia}(p)$ is the Fourier transform of the tensor field with vector index i and spin index $a = 1, 2, 3$. This field describes the collective Bose excitations of the two-dimensional system. $\mathbf{H}$ is the magnetic field, $\xi = c_F(k - k_F)$; c_F is the velocity on the Fermi surface, k_F is the Fermi momentum, μ is the magnetic moment of the quasiparticle, $\mathbf{n}_i = \mathbf{k}_i/k$ is a unit vector, $\beta = T^{-1}$, $\omega = (2n+1)\pi/\beta$ is the Fermi frequency, $p = (k, \omega)$ is the four-momentum, σ_a $(a = 1, 2, 3)$ are the Pauli matrices, $V = S$ is the two-dimensional volume of the system, Z is a normalization constant, g_0 is a negative constant. The momenta k are located in the layer $|k - k_F| < k_0 \ll k_F$.

As it was shown in section 7, Bose condensation is impossible in two-dimensional systems at nonzero temperatures. Nevertheless, superfluidity is possible without Bose condensation, and it is connected with 'long-range correlations' decreasing not exponentially, but more slowly (at $T \neq 0$) (see section 7 and Berezinsky, 1970, 1971; Popov, 1972*b*; Kosterlitz & Thouless, 1973). Furthermore, a number of results obtained under the naive assumption of the existence of the Bose condensate are still valid in spite of the fact that the Bose condensate is actually 'smeared out' by the long-wave fluctuations. Using this naive hypothesis, we can find for the system in question the phase-transition temperature T_c and investigate possible superfluid phases. If $T = 0$ the Bose condensate does really exist, and the naive theory becomes rigorous.

We consider a system described by the functional (17.1), first at $|T - T_c| \ll T_c$ (in the Ginsburg–Landau region). In this region, expanding S_{eff} in powers of c_{ia} and c_{ia}^* and limiting ourselves to terms of the second and fourth orders we get.

$$\begin{aligned} S_{eff} = {} & \sum_p A_{ij}(p) c_{ia}^*(p) c_{ja}(p) - \frac{7\zeta(3) Z^2 \mu^2 H^2 k_F}{4\pi^3 T^2} \sum_p c_{i3}^*(p) c_{i3}(p) \\ & - \frac{7\zeta(3) Z^4 k_F}{16\pi^3 c_F T^3 \beta V} \sum_{p_1 + p_2 = p_3 + p_4} [2c_{ia}^*(p_1) c_{jb}^*(p_2) c_{ia}(p_3) c_{jb}(p_4) \\ & + 2c_{ia}^*(p_1) c_{jb}^*(p_2) c_{ia}(p_3) c_{jb}(p_4) + 2c_{ia}^*(p_1) c_{jb}^*(p_2) c_{ja}(p_3) c_{ib}(p_4) \\ & - 2c_{ia}^*(p_1) c_{ja}^*(p_2) c_{ib}(p_3) c_{jb}(p_4) - c_{ia}^*(p_1) c_{ia}^*(p_2) c_{jb}(p_3) c_{jb}(p_4)], \end{aligned} \tag{17.3}$$

where

$$A_{ij}(p) = \delta_{ij} g_0^{-1} + \frac{4Z^2}{\beta V} \sum_{p_1 + p_2 = p} n_{1i} n_{1j} (i\omega_1 - \xi_1)^{-1} (i\omega_2 - \xi_2)^{-1}. \tag{17.4}$$

Here we suppose **H** to be orthogonal to the film.

We find the phase-transition temperature T_c by equating $A_{ij}(0)$ to zero,

$$A_{ij}(0) = \delta_{ij} g_0^{-1} + \frac{4Z^2}{\beta V} \sum_p n_i n_j (\omega^2 + \xi^2)^{-1} = 0. \tag{17.5}$$

Calculating the sum over frequencies, we rewrite (17.5) in the form

$$g_0^{-1} + \frac{Z^2 k_F}{\pi c_F} \int_0^{c_F k_0} \frac{d\xi}{\xi} \tanh \frac{\beta}{2} \xi = g_0^{-1} + \frac{Z^2 k_F}{\pi c_F} \left(C + \ln \frac{2 c_F \beta k_0}{\pi} \right), \tag{17.6}$$

where the integral depends logarithmically on k_0. Therefore g_0^{-1} should also depend logarithmically on k_0:

$$g^{-1} = g_0^{-1} + \frac{Z^2 k_F}{\pi c_F} \ln \frac{k_F}{k_0}, \tag{17.7}$$

where g_0 no longer depends on k_0. This leads to a formula for T_c:

$$T_c = \frac{2\gamma c_F k_F}{\pi} \exp\left(- \frac{\pi c_F}{Z^2 k_F |g|} \right), \tag{17.8}$$

where $\gamma = e^C$, and C is the Euler constant. We now consider the alternatives for a condensate function at $T < T_c$. Substituting

$$c_{ia}(p) = c_{ia}^{(0)}(p) = (\beta V)^{1/2} \delta_{p0} b_{ia} \tag{17.9}$$

in (17.3) and then making the substitution

$$b_{ia} = 4\left(\frac{T_c \Delta T}{7\zeta(3)} \right)^{1/2} a_{ia} \tag{17.10}$$

we get

$$S_{eff} = -\beta V \frac{16\pi^2 T_c \Delta T}{7\zeta(3)} \Pi, \tag{17.11}$$

where

$$\begin{aligned} \Pi = & -\mathrm{tr}\, AA^+ + \nu\, \mathrm{tr}\, A^+AP + (\mathrm{tr}\, AA^+)^2 + \mathrm{tr}\, AA^+AA^+ \\ & + \mathrm{tr}\, AA^+A^*A^T - \mathrm{tr}\, AA^TA^*A^+ - \tfrac{1}{2} \mathrm{tr}\, AA^T\, \mathrm{tr}\, A^+A^*, \end{aligned} \tag{17.12}$$

where

$$\nu = \frac{7\zeta(3)\mu^2 H^2}{4\pi^2 T_c \Delta T}. \tag{17.13}$$

$$P = \begin{pmatrix} 0 & 0 & 0 \\ 0 & 0 & 0 \\ 0 & 0 & 1 \end{pmatrix} \tag{17.14}$$

is the projection operator on the third axis. Formally, the form (17.12) is the same as that for the three-dimensional system. The main difference is that A is now a 2×3 complex matrix. It corresponds to $2 \times 6 = 12$ modes of collective excitations instead of the 18 in three-dimensional ^{3}He-like systems.

Minimizing Π we obtain the following equation for the condensate matrix A:

$$\begin{aligned} & -A + \nu AP + 2(\mathrm{tr}\, AA^+)A + 2AA^+A + 2A^*A^TA \\ & -2AA^TA^* - A^*\, \mathrm{tr}\, AA^T = 0. \end{aligned} \tag{17.15}$$

This equation has several solutions corresponding to different superfluid phases. We consider the following alternatives:

$$A_1=\frac{1}{2}\begin{pmatrix}1&0&0\\ \mathrm{i}&0&0\end{pmatrix},\quad A_2=\frac{1}{2}\begin{pmatrix}1&0&0\\ 0&1&0\end{pmatrix},\quad A_3=\frac{1}{4}\begin{pmatrix}1&\mathrm{i}&0\\ \mathrm{i}&-1&0\end{pmatrix},$$

$$A_4=\frac{1}{3^{1/2}}\begin{pmatrix}1&0&0\\ 0&0&0\end{pmatrix},\quad A_5=\frac{1}{3^{1/2}}\begin{pmatrix}0&0&0\\ 0&1&0\end{pmatrix},\quad A_6=\left(\frac{1-\nu}{3}\right)^{1/2}\begin{pmatrix}0&0&1\\ 0&0&0\end{pmatrix},$$

$$A_7=\left(\frac{1-\nu}{4}\right)^{1/2}\begin{pmatrix}0&0&1\\ 0&0&\mathrm{i}\end{pmatrix}. \tag{17.16}$$

The corresponding solutions for Π are equal to

$$\Pi_1=-1/4,\quad \Pi_2=-1/4,\quad \Pi_3=-1/8,\quad \Pi_4=\Pi_5=-1/6,$$
$$\Pi_6=-(1-\nu)^2/6,\quad \Pi_7=-(1-\nu)^2/4. \tag{17.17}$$

For the first five phases, the quantity Π does not depend on H. The minimum value of $\Pi=-1/4$ is reached for phases with matrices A_1 and A_2. The phase with the matrix

$$A_1=\frac{1}{2}\begin{pmatrix}1&0&0\\ \mathrm{i}&0&0\end{pmatrix} \tag{17.18}$$

is called the a-phase, and the phase with the matrix

$$A_2=\frac{1}{2}\begin{pmatrix}1&0&0\\ 0&1&0\end{pmatrix} \tag{17.19}$$

is called the b-phase. The a-phase was considered by Cross and Stein (Stein & Cross, 1979).

We calculate the second variation $\delta^2\Pi$. If it is non-negative then the corresponding phase is stable relative to small fluctuations. Knowledge of the quadratic form $\delta^2\Pi$ allows us to determine the phonon variables and find the change in the number of phonon variables upon switching on the magnetic field.

For the a-phase, $\delta^2\Pi$ has the form

$$\begin{aligned}\delta^2\Pi&=\nu(u_{13}^2+u_{23}^2+v_{13}^2+v_{23}^2)\\ &\quad+\tfrac{1}{2}[2(u_{11}+v_{21})^2+(u_{11}-v_{21})^2+(u_{21}+v_{11})^2]\\ &\quad+\tfrac{1}{2}[2(u_{22}-v_{12})^2+(u_{12}-v_{22})^2+(u_{22}+v_{12})^2]\\ &\quad+\tfrac{1}{2}[2(u_{23}-v_{13})^2+(u_{13}-v_{23})^2+(u_{23}+v_{13})^2],\end{aligned} \tag{17.20}$$

where u_{ia} and v_{ia} are the real and imaginary parts of a_{ia}. It follows from

(17.20) that the phonon variables in the a-phase are

$$\begin{aligned} &u_{21}-v_{11}, u_{12}+v_{22}, u_{13}+v_{23} \quad \text{for} \quad \mathbf{H}=0 \quad (\nu=0) \\ &u_{21}-v_{11}, u_{12}+v_{22} \quad \text{for} \quad \mathbf{H}\neq 0 \quad (\nu>0). \end{aligned} \tag{17.21}$$

This means that in the a-phase at $\mathbf{H}=0$ there exist three phonon (Goldstone) modes, while the remaining nine branches of the Bose spectrum are nonphonon (they have a gap as $k\to 0$). Upon switching on the magnetic field, the phonon branch $u_{13}+v_{23}$ acquires a gap.

The calculation of $\delta^2\Pi$ for the b-phase yields

$$\begin{aligned} \delta^2\Pi = \nu(u_{13}^2+u_{23}^2)+(\nu+2)(v_{13}^2+v_{23}^2)+\tfrac{1}{2}[3u_{11}^2+3u_{22}^2 \\ +2u_{11}u_{22}+(u_{12}+u_{21})^2+(v_{11}-v_{22})^2+3v_{12}^2+3v_{21}^2-v_{12}v_{21}]. \end{aligned} \tag{17.22}$$

This expression shows that in the b-phase the phonon variables are

$$\begin{aligned} &u_{12}-u_{21}, v_{11}+v_{22}, u_{13}, u_{23} \quad \text{for} \quad \mathbf{H}=0 \quad (\nu=0) \\ &u_{12}-u_{21}, v_{11}+v_{22} \quad \text{for} \quad \mathbf{H}\neq 0 \quad (\nu>0). \end{aligned} \tag{17.23}$$

In the b-phase at $\mathbf{H}=0$ there exist four phonon (Goldstone) and eight nonphonon modes. Upon switching on the magnetic field, the branches u_{13} and u_{23} acquire gaps and become nonphonon.

Formulas (17.20) and (17.22) also demonstrate the stability of the phases a and b relative to small fluctuations.

In the model considered, both phases a and b have equal free energies, which does not permit us to give preference to one of them over the other. As is seen from (17.21) and (17.23), the phonon variables in the two phases are essentially different. Calculation of the Bose spectrum also gives results that are significantly different for the a- and b-phases.

This calculation may be done using the same methods as for obtaining the Bose spectrum in three-dimensional systems. In order to obtain the Bose spectrum of the system at $T=0$, let us perform the shift transformation $c_{ia}(p)\to c_{ia}(p)+c_{ia}^{(0)}(p)$ in S_{eff}, where $c_{ia}^{(0)}(p)$ is the condensate wavefunction. We can find the Bose spectrum from the quadratic form of S_{eff} (after the shift):

$$Q=\sum_p A_{ijab}(p)c_{ia}^*(p)c_{jb}(p)+\frac{1}{2}\sum_p B_{ijab}(p)[c_{ia}(p)c_{jb}(-p)+c_{ia}^*(p)c_{jb}^*(-p)]. \tag{17.24}$$

The equation for the spectrum is $\det Q=0$. The coefficient functions $A_{ijqb}(p)$, $B_{ijab}(p)$ (integrals of products of two fermion propagators) can be calculated using the methods developed in the previous sections.

We report the results for the condensate wave function $c_{ia}^{(0)}(p)$ and for the Bose spectrum in the a-phase

$$c_{ia}^{(0)}(p) = c(\beta V)^{1/2}\delta_{p0}\delta_{a1}(\delta_{i1} + i\delta_{i2}), \tag{17.25}$$

$$E^2 = \tfrac{1}{2}c_F^2k^2\left(1 - \frac{5c_F^2k^2}{48\Delta^2}\right),$$

$$u_{21} - v_{11},\ u_{12} + v_{22},\ u_{13} + v_{23};\quad \text{(three modes)}$$

$$E^2 = 2\Delta^2 + c_F^2k^2/2$$

$$u_{21} + v_{11}, u_{12} - v_{22}, u_{13} - v_{23}, u_{11} - v_{21}, u_{22} + v_{12},$$

$$u_{23} + v_{13};\quad \text{(six modes)}$$

$$E^2 = 4\Delta^2 + (0.500 - i0.433)c_F^2k^2$$

$$u_{11} + v_{21},\ u_{22} - v_{12},\ u_{23} - v_{13};\quad \text{(three modes).} \tag{17.26}$$

The value of c is connected with the gap Δ of the Fermi spectrum at $T = 0$ and with the transition temperature T_c by the formula

$$2cZ = \Delta = \pi T_c/\gamma, \tag{17.27}$$

where $\gamma = e^C$, and C is the Euler constant.

The Bose spectrum of the a-phase is three-fold degenerate. It contains three phonon modes which are stable under decay into two or more excitations of phonon type. There are also six modes with $\Omega = \Delta 2^{1/2}$ and three modes with $\Omega = 2\Delta$. The last three modes may decay into two fermions and have complex dispersion coefficients. The complex coefficient

$$z = 0.500 - i0.433 \tag{17.28}$$

is the root of the equation

$$\int_0^1 dx\left(\frac{x^2 - z}{1 - x^2}\right)^{1/2} = 0. \tag{17.29}$$

The corresponding results for the b-phase may be stated as follows:

$$c_{ia}^{(0)}(p) = c(\beta V)^{1/2}\delta_{p0}\delta_{ia}, \tag{17.30}$$

$$E^2 = \tfrac{1}{2}c_F^2k^2\left(1 - \frac{5c_F^2k^2}{48\Delta^2}\right),\quad u_{12} - u_{21},\ v_{11} + v_{22};$$

$$E^2 = \tfrac{3}{4}c_F^2k^2\left(1 - \frac{c_F^2k^2}{72\Delta^2}\right),\quad u_{13};$$

$$E^2 = \tfrac{1}{4}c_F^2k^2\left(1 - \frac{c_F^2k^2}{48\Delta^2}\right), \quad u_{23};$$

$$E^2 = 2\Delta^2 + c_F^2k^2/2, \quad u_{11} - u_{22}, u_{12} + u_{21}, v_{11} - v_{22}, v_{12} + v_{21};$$

$$E^2 = 4\Delta^2 + (0.500 - \mathrm{i}0.433)c_F^2k^2, \quad u_{11} + u_{22}, v_{12} - v_{21};$$

$$E^2 = 4\Delta^2 + (0.152 - \mathrm{i}0.218)c_F^2k^2, \quad v_{13};$$

$$E^2 = 4\Delta^2 + (0.849 - \mathrm{i}0.216)c_F^2k^2, \quad v_{23}. \tag{17.31}$$

Here

$$\begin{aligned} 0.500 - \mathrm{i}0.433 &= z_1, \\ 0.152 - \mathrm{i}0.218 &= z_2, \\ 0.849 - \mathrm{i}0.216 &= z_3 \end{aligned} \tag{17.32}$$

are the roots of the equations

$$\int_0^1 \mathrm{d}x\left(\frac{x^2 - z}{1 - x^2}\right)^{1/2}\begin{pmatrix} 1 \\ x^2 \\ 1 - x^2 \end{pmatrix} = 0. \tag{17.33}$$

The constant c obeys the same equation (17.27) which is valid for both a- and b-phases. There are four phonon (stable) modes in the b-phase, four modes with $\Omega = \Delta 2^{1/2}$ and four modes with $\Omega = 2\Delta$. The last four modes are the analogues of pair-breaking modes in the three-dimensional B-phase of ^{3}He. They all have complex dispersion coefficients and may decay into two fermions.

One can see that the Bose spectrum in the a-phase differs from that in the b-phase. All Goldstone modes are stable under decays including the branches of two-dimensional Bogoliubov sound $E = c_F k/2^{1/2}$.

The free energies are the same for the a- and b-phases of the model. This is why we cannot prefer one of them to the other. The Fermi spectrum in both phases has the same form

$$E = (\xi^2 + \Delta^2)^{1/2}. \tag{17.34}$$

Part IV

Crystals, heavy atoms, model Hamiltonians

18

Functional integral approach to the theory of crystals

In this section we will develop an approach to the microscopic theory of periodic structures in the framework of the functional integration method. This approach was suggested by Kapitonov & Popov (1981) and was developed by Andrianov, Kapitonov & Popov (1982, 1983).

Our starting point will be a system of electrons and ions with the Coulomb interaction. The properties of crystals are determined by the collective excitations (phonons). Clearly, a microscopic theory must describe phonons and their interactions starting from the system of electrons and ions. The functional integral method allows us to realize this aim. The main idea is to go from the initial action of electrons and ions to the effective action functional in terms of the electric potential field $\varphi(\mathbf{x}, \tau)$. This field has an immediate physical meaning and provides the collective variable we need.

We can find the static field $\varphi_0(\mathbf{x})$, corresponding to the crystalline structure from the stationary condition for the effective action functional $S_{\mathrm{eff}}[\varphi]$. If $\varphi_0(\mathbf{x})$ is known we can consider small fluctuations in the vicinity of the stationary point of S_{eff}. In order to do this, we have to expand S_{eff} in this neighbourhood and to separate the quadratic form of $\varphi(\mathbf{x}, \tau) - \varphi_0(\mathbf{x})$. It is this quadratic form that defines the spectrum of collective excitations. Forms of the third and higher degrees describe the interaction of these excitations.

The model described below is immediately applicable to the description of metallic hydrogen. If we wish to apply this scheme to other metals we have to modify the starting action in order to take into account the finite sizes of ions defined by their filled electron shells.

We begin with the action functional S for a system of electrons and ions:

$$S = \int_0^\beta \mathrm{d}\tau \int \mathrm{d}^3x \left(\sum_s \bar{\psi}_{\mathrm{e}s}(\mathbf{x}, \tau)\partial_\tau \psi_{\mathrm{e}s}(\mathbf{x}, \tau) + \bar{\psi}_{\mathrm{i}}(\mathbf{x}, \tau)\partial_\tau(\mathbf{x}, \tau)\partial_\tau \psi_{\mathrm{i}}(\mathbf{x}, \tau) \right)$$

$$- \int_0^\beta H'(\tau)\mathrm{d}\tau, \tag{18.1}$$

where

$$H'(\tau) = \int d^3x \sum_s \left(\frac{1}{2m} \nabla\bar{\psi}_{es}(\mathbf{x},\tau)\nabla\psi_{es}(\mathbf{x},\tau) - \lambda_e \bar{\psi}_{es}(\mathbf{x},\tau)\psi_{es}(\mathbf{x},\tau) \right)$$

$$+ \int d^3x \left(\frac{1}{2M} \nabla\bar{\psi}_i(\mathbf{x},\tau)\nabla\psi_i(\mathbf{x},\tau) - \lambda_i \bar{\psi}_i(\mathbf{x},\tau)\psi_i(\mathbf{x},\tau) \right)$$

$$+ \frac{1}{2} \iint d^3x d^3y \frac{e^2}{|\mathbf{x}-\mathbf{y}|} \rho(\mathbf{x},\tau)\rho(\mathbf{y},\tau), \tag{18.2}$$

$$\rho(\mathbf{x},\tau) = \bar{\psi}_i(\mathbf{x},\tau)\psi_i(\mathbf{x},\tau) - \sum_s \bar{\psi}_{es}(\mathbf{x},\tau)\psi_{es}(\mathbf{x},\tau). \tag{18.3}$$

Here $\psi_{es}, \bar{\psi}_{es}$ is the Grassmannian Fermi field of electrons, $s = \pm$ is the spin index. Let us also regard the $\psi_i, \bar{\psi}_i$ field as a Fermi field. We can use the following Fourier expansions of the ψ_{es}, ψ_i fields:

$$\psi_{es}(\mathbf{x},\tau) = (\beta V)^{-1/2} \sum_p e^{i(\omega\tau + \mathbf{kx})} a_{es}(p),$$

$$\psi_i(\mathbf{x},\tau) = (\beta V)^{-1/2} \sum_p e^{i(\omega\tau + \mathbf{kx})} a_i(p), \tag{18.4}$$

where

$$p = (\mathbf{k}, \omega), \quad \omega = (2n+1)\pi/\beta, \quad \beta^{-1} = T.$$

First of all it is convenient to go from the Coulomb interaction to the interaction via the Bose field of electric potential $\varphi(\mathbf{x},\tau)$. This method is also useful in plasma theory as we have seen in section 9. In order to do it let us introduce the following Gaussian integral:

$$\int D\varphi \exp\left[-(8\pi)^{-1} \int d\tau d^3x (\nabla\varphi(\mathbf{x},\tau))^2 \right] \tag{18.5}$$

into the integral with respect to the fields $\psi_{es}, \bar{\psi}_{es}, \psi_i, \bar{\psi}_i$. Performing the shift transformation

$$\varphi(\mathbf{x},\tau) \to \varphi(\mathbf{x},\tau) + ie \int \frac{d^3y \rho(\mathbf{y},\tau)}{|\mathbf{x}-\mathbf{y}|} \tag{18.6}$$

we can cancel the Coulomb interaction term in the action functional (18.1). We then obtain the action functional

$$S[\psi_{es}, \bar{\psi}_{es}, \psi_i, \bar{\psi}_i, \varphi] = -(8\pi)^{-1} \int d\tau d^3x (\nabla\varphi(\mathbf{x},\tau))^2$$

$$+ \int d\tau d^3x \sum_s (\bar{\psi}_{es}(\mathbf{x},\tau)\partial_\tau \psi_{es}(\mathbf{x},\tau) - (2m)^{-1} \nabla\bar{\psi}_{es}(\mathbf{x},\tau)\nabla\psi_{es}(\mathbf{x},\tau)$$

$$+(\lambda_e + ie\varphi(\mathbf{x},\tau))\bar{\psi}_{es}(\mathbf{x},\tau)\psi_{es}(\mathbf{x},\tau))$$

$$+\int d\tau d^3x(\bar{\psi}_i(\mathbf{x},\tau)\partial_\tau\psi_i(\mathbf{x},\tau) - (2M)^{-1}\nabla\bar{\psi}_i(\mathbf{x},\tau)\nabla\psi_i(\mathbf{x},\tau)$$

$$+(\lambda_i - ie\varphi(\mathbf{x},\tau))\bar{\psi}_i(\mathbf{x},\tau)\psi_i(\mathbf{x},\tau)). \tag{18.7}$$

If our systems is in the crystalline state, the ions (whose mass M is much larger than the electron mass m) are located near the sites of some crystalline lattice. In this case it is better to go from the 'field' description of ions to the 'corpuscular' description. In the language of functional integrals this means that we have to integrate not over the fields $\psi_i(\mathbf{x},\tau)$, $\bar{\psi}_i(\mathbf{x},\tau)$, but over the trajectories of ions oscillating near the lattice points. Here the most suitable tool is the formalism of integration over trajectories in the phase space. The last terms in (18.7) depending on $\psi_i, \bar{\psi}_i$ must be replaced by the following expression:

$$\int_0^\beta d\tau \sum_{\mathbf{l}} [i\mathbf{p}_\mathbf{l}(\tau)\partial_\tau\mathbf{q}_\mathbf{l}(\tau) - (2M)^{-1}\mathbf{p}_\mathbf{l}^2(\tau) + \lambda_i - ie\varphi(\mathbf{q}_\mathbf{l}(\tau),\tau)] \tag{18.8}$$

Here $\mathbf{l}$ ($\mathbf{l}\in L$) is the lattice vector which enumerates the ions. Studying the lattice dynamics we may omit the irrelevant term $\sum_\mathbf{l}\lambda_i$ in (18.8).

Now we may integrate over the electron Fermi fields $\psi_{es}, \bar{\psi}_{es}$. The integral is Gaussian and we can express it via a regularized determinant of the operator acting on ψ_{es}. As a result, after replacing the ion terms by (18.8), integrating over Fermi fields, we obtain in place of (18.7) the expression

$$S[\mathbf{p}_\mathbf{l}, \mathbf{q}_\mathbf{l}, \varphi(\mathbf{x},\tau)]$$

$$= -(8\pi)^{-1}\int d\tau d^3x(\nabla\varphi(\mathbf{x},\tau))^2 + 2\ln\det(\partial_\tau + (2m)^{-1}\nabla^2$$

$$+\lambda_e + ie\varphi(\mathbf{x},\tau))/(\partial_\tau + (2m)^{-1}\nabla^2 + \lambda_e) + \int_0^\beta d\tau\sum_\mathbf{l}[i\mathbf{p}_\mathbf{l}(\tau)\partial_\tau\mathbf{q}_\mathbf{l}(\tau)$$

$$-(2M)^{-1}\mathbf{p}_\mathbf{l}^2(\tau) - ie\varphi(\mathbf{q}_\mathbf{l}(\tau),\tau)]. \tag{18.9}$$

The factor of two before ln det is due to taking electron spin into account.

Now we have to integrate $\exp S[\mathbf{p}_\mathbf{l}, \mathbf{q}_\mathbf{l}, \varphi]$ with respect to the field $\varphi(\mathbf{x},\tau)$, the ion momenta $\mathbf{p}_\mathbf{l}(\tau)$ and their coordinates $\mathbf{q}_\mathbf{l}(\tau)$. It is natural to assume that the main contribution in the integral comes from the fields $\varphi(\mathbf{x},\tau)$, the momenta $\mathbf{p}_\mathbf{l}(\tau)$ and the coordinates $\mathbf{q}_\mathbf{l}(\tau)$ which are close to some equilibrium values. These values can be found from the equations

$$\frac{\nabla^2}{4\pi}\varphi(\mathbf{x},\tau) + ie\left\{2[\partial_\tau + (2m)^{-1}\nabla^2 + \lambda_e + ie\varphi(\mathbf{x},\tau)]^{-1}_{\mathbf{x},\tau;\mathbf{x},\tau} - \sum_{\mathbf{l}}\delta(\mathbf{x}-\mathbf{q}_{\mathbf{l}}(\tau))\right\} = 0,$$

$$i\partial_\tau \mathbf{q}_{\mathbf{l}}(\tau) - M^{-1}\mathbf{p}_{\mathbf{l}}(\tau) = 0,$$

$$-i\partial_\tau \mathbf{p}_{\mathbf{l}}(\tau) + ie\nabla\varphi(\mathbf{q}_{\mathbf{l}}(\tau),\tau) = 0. \tag{18.10}$$

First of all we consider the stationary solution of (18.10) of the form

$$\mathbf{p}_{\mathbf{l}} = 0, \quad \mathbf{q}_{\mathbf{l}} = \mathbf{l} \in L, \quad \varphi = -i\phi(\mathbf{x}), \tag{18.11}$$

where $\mathbf{l} \in L$ is a lattice vector. It is clear that the second equation is fulfilled and the third equation holds because $\nabla\varphi = 0$ in the lattice points. The first equation will hold if $\phi(\mathbf{x})$ obeys the following nonlinear equation of the selfconsistent field:

$$\nabla^2\phi(\mathbf{x}) = 4\pi e\left\{2[\partial_\tau + (2m)^{-1}\nabla^2 + \lambda_e + e\phi(\mathbf{x})]^{-1}_{\mathbf{x},\tau;\mathbf{x},\tau} - \sum_{\mathbf{l}\in L}\delta(\mathbf{x}-\mathbf{l})\right\}. \tag{18.12}$$

We cannot solve (18.12) explicitly. Nevertheless it is not hard to obtain an approximate solution of this equation, if we linearize its right-hand side relative to $\phi(\mathbf{x})$, i.e. take only the first two terms in the functional expansion of ln det in $\phi(\mathbf{x})$. Let us write down the linearized equation, moving the terms depending on $\phi(\mathbf{x})$ to the left-hand side:

$$\nabla^2\phi(\mathbf{x}) + 8\pi e^2\int d\tau' d^3y G_0(\mathbf{x}-\mathbf{y},\tau-\tau')G_0(\mathbf{y}-\mathbf{x},\tau'-\tau)\phi(\mathbf{y})$$

$$= 4\pi e\left\{\rho_0 - \sum_{\mathbf{l}\in L}\delta(\mathbf{x}-\mathbf{l})\right\}. \tag{18.13}$$

The constant ρ_0 on the right-hand side

$$\rho_0 = 2[\partial_\tau + (2m)^{-1}\nabla^2 + \lambda_e]^{-1}_{\mathbf{x},\tau;\mathbf{x},\tau} = 2G_0(0,0) \tag{18.14}$$

is the density of electron gas with chemical potential λ_e. For the system to be electroneutral, we must take $\rho_0 = V_0^{-1}$, where V_0 is the volume of an elementary crystalline cell. G_0 in (18.13) and (18.14) is the Green function of a free electron.

It is not hard to solve (18.13) using the Fourier transform. The solution

$$\phi(\mathbf{x}) = 4\pi e\rho_0 \sum_{0\neq \mathbf{k}\in L^*} \frac{e^{i\mathbf{k}\mathbf{x}}}{k^2\varepsilon(\mathbf{k})} \tag{18.15}$$

has the form of a sum over nonzero vectors of the inverse lattice L^*. Here $\varepsilon(\mathbf{k})$ is defined by the formulae

$$\varepsilon(\mathbf{k}) = 1 - \frac{8\pi e^2}{k^2\beta V}\sum_{\mathbf{k}_1,\omega_1} G_0(\omega_1,\mathbf{k}_1)G_0(\omega_1,\mathbf{k}_1-\mathbf{k}) \tag{18.16}$$

$$= 1 - \frac{8\pi e^2}{k^2 V}\sum_{\mathbf{k}_1}\frac{n(\mathbf{k}_1) - n(\mathbf{k}_1-\mathbf{k})}{\frac{k_1^2}{2m} - \frac{(\mathbf{k}_1-\mathbf{k})^2}{2m}}. \tag{18.17}$$

$\varepsilon(\mathbf{k})$ is the dielectric function in the random phase approximation (RPA). The solution (18.15) of the linearized equation (18.13) must give a good approximation to an exact solution for systems of large density.

Regarding $\phi(\mathbf{x})$ as a known function, we can take into account small deviations of $\varphi(\mathbf{x},\tau)$ from $-i\phi(\mathbf{x})$, of $\mathbf{q}_l(\tau)$ from $\mathbf{l}$ and $\mathbf{p}_l(\tau)$ from zero. We start by considering the first equation in (18.10) as a equation for $\varphi(\mathbf{x},\tau)$ if $\mathbf{q}_l(\tau)$ are given functions. Physically it means that we have to obtain the electric potential for a given low of motion of heavy particles. This is the usual way in the so-called adiabatic approach. We denote

$$\varphi(\mathbf{x},\tau) = -i\phi(\mathbf{x}) - i\delta\phi(\mathbf{x},\tau). \tag{18.18}$$

Substituting (18.18) into the first equation in (18.10) we obtain the equation

$$\begin{aligned}\nabla^2(\phi(\mathbf{x}) + \delta\phi(\mathbf{x},\tau)) &= 4\pi e\Big\{2[2\partial_\tau + (2m)^{-1}\nabla^2 + \lambda_e + e(\phi(\mathbf{x}) + \delta\phi(\mathbf{x},\tau))]^{-1}_{\mathbf{x},\tau;\mathbf{x},\tau} \\ &\quad - \sum_{\mathbf{l}\in L}\delta(\mathbf{x} - \mathbf{q}_l(\tau))\Big\}.\end{aligned} \tag{18.19}$$

Let us subtract (18.12) from (18.19) and linearize relative to $\delta\phi(\mathbf{x},\tau)$. We obtain the following linearized equation for $\delta\phi(\mathbf{x},\tau)$:

$$\begin{aligned}\nabla^2\delta\phi(\mathbf{x},\tau) &+ 8\pi e^2\int d\tau' d^3y G(\mathbf{x},\tau,\mathbf{y},\tau'|\phi)G(\mathbf{y},\tau',\mathbf{x},\tau|\phi)\delta\phi(\mathbf{y},\tau') \\ &= 4\pi e\sum_{\mathbf{l}\in L}[\delta(\mathbf{x}-\mathbf{l}) - \delta(\mathbf{x} - \mathbf{q}_l(\tau))].\end{aligned} \tag{18.20}$$

Here $G(\mathbf{x},\tau,\mathbf{y},\tau'|\phi)$ is the Green function of electron in the periodic potential $\phi(\mathbf{x})$.

Linearization of the right-hand side of (18.20) relative to $\mathbf{q}_l(\tau)-\mathbf{l}$ gives

$$4\pi e\sum_{l\in L}(\mathbf{q}_l(\tau)-l,\nabla\delta(\mathbf{x}-\mathbf{l})). \tag{18.21}$$

The Green function G can be written as

$$G(\mathbf{x},\tau,\mathbf{y},\tau'|\phi) \\ =(\beta V)^{-1}\sum_{\omega,\mathbf{k},\mathbf{k}_1}G(\omega,\mathbf{k},\mathbf{k}_1)\exp i(\omega(\tau-\tau_1)+\mathbf{kx}-\mathbf{k}_1\mathbf{y}). \tag{18.22}$$

Taking into account that the periodic potential $\phi(\mathbf{x})$ has only Fourier components with $\mathbf{k}\in L^*$, we can write

$$\mathbf{k}=\mathbf{p}+\mathbf{K},\quad \mathbf{k}_1=\mathbf{p}+\mathbf{K}_1,\quad \mathbf{K},\mathbf{K}_1\in L^*,\quad \mathbf{p}\in B_1, \tag{18.23}$$

i.e. $\mathbf{k}$ and $\mathbf{k}_1$ can differ only by an inverse lattice vector. $\mathbf{p}\in B_1$ in (18.23) means that $\mathbf{p}$ belongs to the first Brillouin zone B_1. As it is well known, B_1 is a polyedron with facets formed by the planes orthogonal to vectors of the lattice points nearest to zero, and intersecting these vectors at their centres.

So we can rewrite (18.22) as follows:

$$G(\mathbf{x},\tau;\mathbf{y},\tau'|\phi)=(\beta V)^{-1}\sum_{\substack{\omega,\mathbf{p}\in B_1\\ \mathbf{k},\mathbf{k}_1\in L^*}}G(\omega,\mathbf{p},\mathbf{k},\mathbf{k}_1) \\ \times\exp(i\omega(\tau-\tau')+i\mathbf{kx}-i\mathbf{k}_1\mathbf{y}+i\mathbf{p}(\mathbf{x}-\mathbf{y})) \tag{18.24}$$

Using the substitutions

$$\delta\phi(\mathbf{x},\tau)=(\beta V)^{-1}\sum_{\substack{\omega,\mathbf{p}\in B_1\\ \mathbf{k}\in L^*}}\delta\phi(\omega,\mathbf{p}+\mathbf{k})e^{i(\omega\tau+(\mathbf{p}+\mathbf{k},\mathbf{x}))}, \tag{18.25}$$

$$\nabla\delta(\mathbf{x}-\mathbf{l})=iV^{-1}\sum_{\mathbf{p}\in B_1,\mathbf{k}\in L^*}(\mathbf{p}+\mathbf{k})e^{i(\mathbf{p}+\mathbf{k},\mathbf{x}-\mathbf{l})}, \tag{18.26}$$

$$\mathbf{q}_l(\tau)-\mathbf{l}=\beta^{-1}\sum_{\omega}\delta\mathbf{q}_l(\omega)e^{i\omega\tau}, \tag{18.27}$$

we may rewrite (18.20) with the right-hand side (18.21) in terms of Fourier coefficients:

$$-(\mathbf{p}+\mathbf{k})^2\delta\phi(\omega,\mathbf{p}+\mathbf{k})+8\pi e^2\sum_{\mathbf{k}'\in L^*}\Pi_e(\omega;\mathbf{p}+\mathbf{k},\mathbf{p}+\mathbf{k}')\delta\phi(\omega,p+\mathbf{k}') \\ =4\pi ie\sum_{l\in L}(\mathbf{p}+\mathbf{k},\delta\mathbf{q}_l(\omega))e^{-i\mathbf{pl}}. \tag{18.28}$$

Here Π_e is the so-called electron loop in a periodic field, i.e.

$$\Pi_e(\omega, \mathbf{p}+\mathbf{k}, \mathbf{p}+\mathbf{k}') = \frac{2}{\beta V} \sum_{\omega_1, \mathbf{p}_1 \in B_1} G(\omega_1, \mathbf{p}_1, \mathbf{k}) G(\omega + \omega_1, \mathbf{p}+\mathbf{p}_1, \mathbf{k}'). \tag{18.29}$$

Introducing the matrix

$$\varepsilon(\omega, \mathbf{p}+\mathbf{k}, \mathbf{p}+\mathbf{k}') = \delta_{\mathbf{k},\mathbf{k}'} - \frac{8\pi e^2}{|\mathbf{p}+\mathbf{k}'|^2} \Pi_e(\omega, \mathbf{p}+\mathbf{k}', \mathbf{p}+\mathbf{k}) \tag{18.30}$$

we can write the solution of (18.28) as follows:

$$\delta\phi(\omega, \mathbf{p}+\mathbf{k}) = 4\pi e \sum_{\mathbf{l}\in L} (\delta\mathbf{q}_{\mathbf{l}}(\omega), A(\omega, \mathbf{p}+\mathbf{k})) e^{-i\mathbf{p}\mathbf{l}}, \tag{18.31}$$

where

$$A(\omega, \mathbf{p}+\mathbf{k}) = -\mathrm{i} \sum_{\mathbf{k}'\in L^*} \varepsilon^{-1}(\omega, \mathbf{p}+\mathbf{k}, \mathbf{p}+\mathbf{k}') \frac{\mathbf{p}+\mathbf{k}'}{|\mathbf{p}+\mathbf{k}'|^2}. \tag{18.32}$$

If we know $\phi(\mathbf{x})$ and $\delta\phi(\mathbf{x}, \tau)$ we can expand the functional (18.9) into a power series in the deviation of $\varphi(\mathbf{x}, \tau)$ from $-\mathrm{i}(\phi(\mathbf{x}) + \delta\phi(\mathbf{x}, \tau))$, i.e. into a power series in $\varphi_1(\mathbf{x}, \tau)$, where

$$\varphi_1(\mathbf{x}, \tau) = \varphi(\mathbf{x}, \tau) + \mathrm{i}(\phi(\mathbf{x}) + \delta\phi(\mathbf{x}, \tau)). \tag{18.33}$$

There is no linear term in $\varphi_1(\mathbf{x}, \tau)$ in this expansion, because the field $-\mathrm{i}(\phi(\mathbf{x}) + \delta\phi(\mathbf{x}, \tau))$ obeys the 'stationary' equation. The terms which do not depend on $\varphi_1(\mathbf{x}, \tau)$ are followed by the terms of the second order, third order and so on. The coefficient functions of this expansion depend on $\mathbf{q}_{\mathbf{l}}(\tau) - \mathbf{l}$ and can be expanded into a power series in $\mathbf{q}_{\mathbf{l}}(\tau) - \mathbf{l}$.

We shall first consider the terms quadratic in $\varphi_1(\mathbf{x}, \tau)$, $\mathbf{q}_{\mathbf{l}}(\tau) - \mathbf{l}$. These terms describe noninteracting excitations. The third- and higher-order terms describe the interaction of excitations.

The quadratic form of the variable φ_1 can be obtained by putting $\mathbf{q}_{\mathbf{l}}(\tau) = \mathbf{l}$ in the coefficient functions of the above-mentioned expansion. So it is clear that the quadratic form of the variable is nothing but the second variation form of the functional (18.9) relative to the variable $\varphi(\mathbf{x}, \tau)$ at $\varphi(\mathbf{x}, \tau) = -\mathrm{i}\phi(\mathbf{x})$. We can write this form as

$$-(8\pi)^{-1} \int d\tau d^3x (\nabla\varphi_1(\mathbf{x}, \tau))^2$$

$$+ e^2 \int d\tau d\tau' d^3x d^3y \varphi_1(\mathbf{x}, \tau) \varphi_1(\mathbf{y}, \tau') G(\mathbf{x}, \tau; \mathbf{y}, \tau' | \phi) G(\mathbf{y}, \tau'; \mathbf{x}, \tau | \phi), \tag{18.34}$$

where $G(\mathbf{x},\tau;\mathbf{y},\tau'|\phi)$ is the electron Green function in the periodic potential $\phi(\mathbf{x})$. This form describes collective excitations of the system of the plasma oscillation type.

We have to add to (18.34) the quadratic form of $\mathbf{q}_l(\tau)-\mathbf{l}$, $\mathbf{p}_l(\tau)$ describing lattice oscillations. In order to calculate this form, we substitute $\varphi = -i(\phi(\mathbf{x})+\delta(\mathbf{x},\tau))$ into (18.9) and then separate the terms quadratic in $\mathbf{q}_l(\tau)-\mathbf{l}$, $\mathbf{p}_l(\tau)$. First of all we write the last term in (18.9) in the form

$$-e\int_0^\beta d\tau\sum_{\mathbf{l}}[\phi(\mathbf{q}_l(\tau))+\delta\phi(\mathbf{q}_l(\tau),\tau)]$$

$$=-e\int d\tau d^3\mathbf{x}\left(\sum_{\mathbf{l}}\delta(\mathbf{x}-\mathbf{l})\right)(\phi(\mathbf{x})+\delta\phi(\mathbf{x},\tau))$$

$$-e\int d\tau d^3x(\phi(\mathbf{x})+\delta\phi(\mathbf{x},\tau))\left[\sum_{\mathbf{l}}\delta(\mathbf{x}-\mathbf{q}_l(\tau))-\delta(\mathbf{x}-\mathbf{l}))\right]. \qquad (18.35)$$

We combine the first term on the right-hand side of (18.35) with the expression

$$(8\pi)^{-1}\int d\tau d^3x[\nabla(\phi(\mathbf{x})+\delta\phi(\mathbf{x},\tau))]^2$$

$$+2\ln\det[\partial_\tau+(2m)^{-1}\nabla^2+\lambda_e+e(\phi(\mathbf{x})+\delta\phi(\mathbf{x},\tau))]/[\partial_\tau+(2m)^{-1}\nabla^2+\lambda_e]. \qquad (18.36)$$

We are interested in the terms quadratic in $\mathbf{q}_l(\tau)-\mathbf{l}$. These terms are given by the quadratic part of the expansion of (18.36) plus the first term on the right-hand side of (18.35). There are no linear terms in $\delta\phi(\mathbf{x},\tau)$ due to the equation for $\phi(\mathbf{x})$. So we obtain

$$(8\pi)^{-1}\int d\tau d^3x(\nabla\delta\phi(\mathbf{x},\tau))^2-e^2\int d\tau d\tau' d^3x d^3y\,\delta\phi(\mathbf{x},\tau)\delta\phi(\mathbf{y},\tau)$$

$$\times G(\mathbf{x},\tau;\mathbf{y},\tau'|\phi)G(\mathbf{y},\tau';\mathbf{x},\tau|\phi). \qquad (18.37)$$

Now we consider the second term on the right-hand side of (18.35) which can be written as follows:

$$-e\int d\tau\sum_{\mathbf{l}}[\phi(\mathbf{q}_l(\tau))-\phi(\mathbf{l})]+e\int d\tau d^3x\,\delta\phi(\mathbf{x},\tau)(\mathbf{q}_l(\tau)-\mathbf{l},\nabla\delta(\mathbf{x}-\mathbf{l})). \qquad (18.38)$$

In the first term we may make the replacement

$$\phi(\mathbf{q}_l(\tau))-\phi(\mathbf{l})\approx\tfrac{1}{2}(\delta\mathbf{q}_l(\tau),\nabla)^2\phi(\mathbf{x})|_{\mathbf{x}=\mathbf{l}}$$

and then rewrite the first term in (18.38) as

$$\frac{e}{2\beta}\sum_{\substack{\omega,\mathbf{k}\in L^*\\ \mathbf{l}\in L}}(\delta\mathbf{q}_{\mathbf{l}}(\omega))_i(\delta\mathbf{q}_{\mathbf{l}}(-\omega))_j K_i K_j \phi(\mathbf{K}). \tag{18.39}$$

Now we use the equation (18.20) for $\delta\phi(\mathbf{x},\tau)$ and rewrite (18.37) as

$$-\frac{e}{2}\int \mathrm{d}\tau \mathrm{d}^3x \sum_{\mathbf{l}\in L}(\delta\mathbf{q}_{\mathbf{l}}(\tau), \nabla\delta(\mathbf{x}-\mathbf{l}))\delta\phi(\mathbf{x},\tau). \tag{18.40}$$

Combining this term with the second added in (18.38) we get

$$\frac{e}{2}\int \mathrm{d}\tau \mathrm{d}^3x \sum_{\mathbf{l}\in L}(\delta\mathbf{q}_{\mathbf{l}}(\tau), \nabla\delta(\mathbf{x}-\mathbf{l}))\delta\phi(\mathbf{x},\tau)$$

$$= -\frac{e}{2}\int \mathrm{d}\tau \sum_{\mathbf{l}\in L}(\delta\mathbf{q}_{\mathbf{l}}(\tau), \nabla\delta\phi(\mathbf{x},\tau))|_{\mathbf{x}=\mathbf{l}}. \tag{18.41}$$

Replacing $\delta\phi(\mathbf{x},\tau)$ in (18.41) by its expression according to (18.25), (18.31) and (18.32), we obtain

$$-\frac{2\pi e^2}{\beta}\sum_{\omega}\sum_{\mathbf{l}_1\neq\mathbf{l}_2}(\delta\mathbf{q}_{\mathbf{l}_1}(\omega))_i(\delta\mathbf{q}_{\mathbf{l}_2}(-\omega))_j a_{ij}(\omega,\mathbf{l}_1-\mathbf{l}_2), \tag{18.42}$$

where

$$a_{ij}(\omega,\mathbf{l}_1-\mathbf{l}_2)$$
$$= V^{-1}\sum_{\mathbf{p}\in B_1} e^{i(\mathbf{p},\mathbf{l}_1-\mathbf{l}_2)}\sum_{\mathbf{K},\mathbf{K}'\in L^*}(\mathbf{p}+\mathbf{K})_i(\mathbf{p}+\mathbf{K}')_j\varepsilon^{-1}(\omega,\mathbf{p}+\mathbf{K},\mathbf{p}+\mathbf{K}')|\mathbf{p}+\mathbf{K}'|^{-2}. \tag{18.43}$$

We thus derive the following quadratic form of $\delta\mathbf{q}_{\mathbf{l}}, \mathbf{p}_{\mathbf{l}}$:

$$\beta^{-1}\sum_{\omega}\Bigg[\sum_{\mathbf{l}}(\omega\mathbf{p}_{\mathbf{l}}(\omega)\delta\mathbf{q}_{\mathbf{l}}(\omega)-(2M)^{-1}\mathbf{p}_{\mathbf{l}}(\omega)\mathbf{p}_{-\mathbf{l}}(-\omega))$$
$$+\frac{e}{2}\sum_{\mathbf{l}}(\delta\mathbf{q}_{\mathbf{l}}(\omega))_i(\delta\mathbf{q}_{\mathbf{l}}(-\omega))_j\sum_{\mathbf{K}\in L^*}K_iK_j\phi(\mathbf{K})$$
$$-2\pi e^2\sum_{\mathbf{l}_1,\mathbf{l}_2}(\delta\mathbf{q}_{\mathbf{l}_1}(\omega))_i(\delta\mathbf{q}_{\mathbf{l}_2}(-\omega))_j a_{ij}(\omega,\mathbf{l}_1-\mathbf{l}_2)\Bigg]. \tag{18.44}$$

It is this form that describes harmonic oscillations of the lattice.

Let us pass from $\mathbf{p}_l$, $\delta\mathbf{q}_l$ to the new (normal) variables:

$$\delta\mathbf{q}_l(\omega) = N^{-1/2} \sum_{\mathbf{p}\in B_1} \mathbf{q}_\mathbf{p}(\omega) e^{i\mathbf{p}l},$$

$$\mathbf{p}_l(\omega) = N^{-1/2} \sum_{\mathbf{p}\in B_1} \mathbf{p}_\mathbf{p}(\omega) e^{i\mathbf{p}l}. \tag{18.45}$$

This is a standard procedure for investigating lattice vibrations. The quadratic form (18.44) becomes simpler in the new variables:

$$\beta^{-1} \sum_{\omega,\mathbf{p}\in B_1} [\omega \mathbf{p}_\mathbf{p}(\omega)\mathbf{q}_{-\mathbf{p}}(-\omega) - (2M)^{-1}\mathbf{p}_\mathbf{p}(\omega)\mathbf{p}_{-\mathbf{p}}(-\omega)$$
$$- \tfrac{1}{2} D_{ij}(\omega,\mathbf{p})(\mathbf{q}_\mathbf{p}(\omega))_i(\mathbf{q}_{-\mathbf{p}}(-\omega))_j]. \tag{18.46}$$

It is usual to call the matrix $D_{ij}(\omega, \mathbf{p})$ in (18.46) the dynamical matrix. Its eigenvalues determine the spectrum of lattice variations. We can write this matrix explicitly as

$$D_{ij}(\omega,\mathbf{p}) = 4\pi e^2 \rho_0 \sum_{\mathbf{K},\mathbf{K}'\in L^*} \frac{(\mathbf{p}+\mathbf{K})_i(\mathbf{p}+\mathbf{K}')_j}{|\mathbf{p}+\mathbf{K}'|^2} \varepsilon^{-1}(\omega, \mathbf{p}+\mathbf{K}, \mathbf{p}+\mathbf{K}')$$
$$- e \sum_{\mathbf{K}\in L^*} K_i K_j \phi(\mathbf{K}). \tag{18.47}$$

In the first approximation $\phi(\mathbf{k})$ is defined by (18.15):

$$\phi(\mathbf{K}) = 4\pi e \rho_0 K^{-2} \varepsilon^{-1}(\mathbf{K}). \tag{18.48}$$

The same approximation for ε^{-1} is

$$\varepsilon^{-1}(\omega, \mathbf{p}+\mathbf{K}, \mathbf{p}+\mathbf{K}') = \delta_{\mathbf{K}\mathbf{K}'} \varepsilon^{-1}(\omega, \mathbf{p}+\mathbf{K}), \tag{18.49}$$

$$\varepsilon(\omega, \mathbf{K}) = 1 - \frac{8\pi e^2}{K^2 \beta V} \sum_{\mathbf{K}_1,\omega_1} G_0(\omega_1, \mathbf{K}_1) G_0(\omega+\omega_1, \mathbf{K}+\mathbf{K}_1)$$
$$= 1 - \frac{8\pi e^2}{K^2 V} \sum_{\mathbf{K}_1} \frac{n(\mathbf{K}_1) - n(\mathbf{K}+\mathbf{K}_1)}{i\omega + \dfrac{k_1^2}{2m} - \dfrac{(\mathbf{K}+\mathbf{K}_1)^2}{2m}} \tag{18.50}$$

is the dielectric function in RPA.

Using approximations (18.48) and (18.49) we may rewrite the dynamical matrix as

$$D_{ij}(\omega,\mathbf{p}) = M\omega_\mathbf{p}^2 \sum_{\mathbf{K}\in L^*} \left[\frac{(\mathbf{p}+\mathbf{K})_i(\mathbf{p}+\mathbf{K})_j}{(\mathbf{p}+\mathbf{K})^2 \varepsilon(\omega, \mathbf{p}+\mathbf{K})} - \frac{K_i K_j}{K^2 \varepsilon(\mathbf{K})} \right], \tag{18.51}$$

where

$$\omega_p^2 = \frac{4\pi e^2 \rho}{M} \tag{18.52}$$

is a square of the so-called ion-plasma frequency. Relation (18.51) implies that the dynamical matrix $D_{ij}(\omega, \mathbf{p})$ has three real positive eigenvalues

$$D\mathbf{e}_{\mathbf{p},\lambda} = M\omega_{\mathbf{p},\lambda}^2 \mathbf{e}_{\mathbf{p},\lambda}; \quad \lambda = 1, 2, 3. \tag{18.53}$$

Here $\mathbf{e}_{\mathbf{p},\lambda}$ are orthonormal eigenvectors of D_{ij} and $\omega_{\mathbf{p},\lambda}$ are the eigenfrequencies.

We see that the functional integral approach allows us to describe a crystalline state in a system of electrons and ions and to go from the initial variables of functional integration to some new variables (18.45) corresponding to normal oscillations of the crystalline lattice. In the next section we apply this approach to construct the effective interaction between electrons near the Fermi surface. It is not hard to show that this interaction for the model considered has an attractive character and may lead to superconductivity at sufficiently low temperatures.

19

Effective interaction of electrons near the Fermi surface

As we know, the superconductivity phenomenon exists due to attractive interaction between electrons in the narrow energetic layer near the Fermi surface. The effective interaction of electrons consists of two parts: the Coulomb repulsive potential and some attractive potential arising from the phonon exchange. In this section we will derive this effective interaction potential in the framework of the functional integral method. Here the procedure of the previous section will be slightly changed. At the first stage we integrate over the electron fields with momenta outside the narrow layer near the Fermi surface. Then we integrate over the electric potential field and also over ion trajectories in the phase space. So we come down to the effective interaction potential between electrons near the Fermi surface.

The starting action functional is

$$S[\psi_e, \bar{\psi}_e, \mathbf{p}_l, \mathbf{q}_l, \varphi] = -(8\pi)^{-1} \int d\tau d^3x (\nabla\varphi(\mathbf{x}, \tau))^2$$
$$+ \int d\tau d^3x \sum_s \bar{\psi}_{es}(\mathbf{x}, \tau)(\partial_\tau + (2m)^{-1}\nabla^2 + \lambda_e + ie\varphi(\mathbf{x}, \tau))\psi_{es}(\mathbf{x}, \tau)$$
$$+ \int d\tau \sum_l [i\mathbf{p}_l(\tau)\partial_\tau \mathbf{q}_l(\tau) - (2M)^{-1}\mathbf{p}_l^2(\tau) + \lambda_i - ie\varphi(\mathbf{q}_l(\tau), \tau)]. \tag{19.1}$$

At the first step of functional interaction we extract the slow varying component in the electric potential field $\varphi(\mathbf{x}, \tau)$ by performing the following shift transformation:

$$\varphi(\mathbf{x}, \tau) = \tilde{\varphi}(\mathbf{x}, \tau) - i\phi(\mathbf{x} | \mathbf{q}_l(\tau)). \tag{19.2}$$

The equation for $\phi(\mathbf{x} | \mathbf{q}_l(\tau))$ can be obtained from the variation principle after integration with respect to the field $\psi_{1es}(\mathbf{x}, \tau)$, and the fields $\psi_{1es}(\mathbf{x}, \tau)$, $\bar{\psi}_{1es}(\mathbf{x}, \tau)$ of electrons outside the shell near the Fermi surface. The fields ψ_0, ψ_1 are defined by the formulae

$$\psi_{0es}(\mathbf{x}, \tau) = (\beta V)^{-1/2} \sum_{\substack{\mathbf{k},\omega \\ |k - k_F| < k_0}} e^{i(\omega\tau + \mathbf{kx})} \psi_{es}(\mathbf{k}, \omega),$$

$$\psi_{1es}(\mathbf{x},\tau) = (\beta V)^{-1/2} \sum_{\substack{\mathbf{k},\omega \\ |k-k_F|>k_0}} e^{i(\omega\tau + \mathbf{kx})} \psi_{es}(\mathbf{k},\omega),$$

$$\psi_{es}(\mathbf{x},\tau) = \psi_{0es}(\mathbf{x},\tau) + \psi_{1es}(\mathbf{x},\tau). \tag{19.3}$$

Upon integrating $\exp S_0$ with respect to the fields $\tilde{\varphi}(\mathbf{x},\tau)$, $\psi_{1es}(\mathbf{x},\tau)$, $\bar{\psi}_{1es}(\mathbf{x},\tau)$ we obtain

$$\int \mathrm{D}\tilde{\varphi}\mathrm{D}\bar{\psi}_1 \mathrm{D}\psi_1 \exp S_0[\psi_0 + \psi_1, \bar{\psi}_0 + \bar{\psi}_1, \mathbf{p}_l, \mathbf{q}_l, \tilde{\varphi} - \mathrm{i}\phi]$$
$$= \exp S_{\mathrm{eff}}[\psi_0, \bar{\psi}_0, \mathbf{p}_l, \mathbf{q}_l, \phi]. \tag{19.4}$$

In the first approximation S_{eff} is a sum of three terms

$$S_{\mathrm{eff}} = S_{\mathrm{el}}[\psi_0, \bar{\psi}_0] + S_{\mathrm{ph}}[\mathbf{p}_l, \mathbf{q}_l] + S_{\mathrm{el\text{-}ph}}. \tag{19.5}$$

Here $S_{\mathrm{el}}[\psi_0, \bar{\psi}_0]$ is the action functional describing electrons on the shell and their pair Coulomb interaction. It consists of terms of the second and fourth order in $\psi_0, \bar{\psi}_0$. The term $S_{\mathrm{ph}}[\mathbf{p}_l, \mathbf{q}_l]$ describes a phonon system, $S_{\mathrm{el\text{-}ph}}$ is the electron–phonon interaction.

Formally S_{eff} also contains terms of the sixth and higher degrees on $\psi_0, \bar{\psi}_0$. If the shell is sufficiently narrow, such terms can be neglected.

It is not hard to write down expressions for each term in (19.5). The first term is

$$S_{\mathrm{el}} = \sum_{\omega,s} \int \mathrm{d}^3x \bar{\psi}_{0s}(\mathbf{x},\omega)(\mathrm{i}\omega - (2m)^{-1}\nabla^2 + \lambda)\psi_{0s}(\mathbf{x},\omega)$$
$$- \sum_{\omega,s} \int \mathrm{d}^3x\,\mathrm{d}^3y \bar{\psi}_{0s}(\mathbf{x},\omega)\,\Sigma_0(\mathbf{x},\mathbf{y},\omega)\psi_{0s}(\mathbf{y},\omega)$$
$$- (2\beta)^{-1} \sum_{\omega} \int \mathrm{d}^3x\,\mathrm{d}^3y \frac{e^2}{|\mathbf{x}-\mathbf{y}|} \varepsilon^{-1}(\mathbf{x},\mathbf{y},\omega)\rho_0(\mathbf{x},\omega)\rho_0(\mathbf{y},-\omega), \tag{19.6}$$

where

$$\rho_0(\mathbf{x},\omega) = \sum_{s,\omega_1} \bar{\psi}_{0s}(\mathbf{x},\omega_1)\psi_{0s}(\mathbf{x},\omega+\omega_1). \tag{19.7}$$

Here $s = \pm$ is a spin index, $\Sigma_0(\mathbf{x},\mathbf{y},\omega)$ is the self-energy part, $\varepsilon(\mathbf{x},\mathbf{y},\omega)$ is a dielectric function.

The slow part of the electric potential field ϕ can be written as

$$\phi = \phi_0(\mathbf{x}) + \delta\phi(\mathbf{x}|\mathbf{q}_l(\tau)). \tag{19.8}$$

In the first approximation we have

$$\phi_0(\mathbf{x}) = 4\pi e\rho \sum_{0\neq \mathbf{k}\in L^*} \frac{e^{i\mathbf{k}\mathbf{x}}}{k^2\varepsilon(\mathbf{k})}. \tag{19.9}$$

Here $\mathbf{k}$ stands for the inverse lattice vectors, $\varepsilon(\mathbf{k})$ is a static dielectric function of the electron gas. The term $\delta\phi$ in (19.8) is defined by the formulae

$$\delta\phi(\mathbf{x}, \delta\mathbf{q}_l(\tau)) = (\beta V)^{-1} \sum_{\substack{\omega,\mathbf{p}\in B_1,\\ \mathbf{k}\in L^*}} e^{i(\omega\tau(\mathbf{p}+\mathbf{k},\mathbf{x}))}\delta\phi(\omega, \mathbf{p}+\mathbf{k}),$$

$$\delta\phi(\omega, \mathbf{p}+\mathbf{k}) = -4\pi i e N^{1/2}\left(\delta\mathbf{q}_l(\omega), \frac{\mathbf{p}+\mathbf{k}}{(\mathbf{p}+\mathbf{k})^2}\varepsilon^{-1}(\omega, \mathbf{p}+\mathbf{k})\right),$$

$$\mathbf{q}_p(\omega) = N^{-1/2}\sum_{l\in L} \delta\mathbf{q}_l(\omega)e^{-ipl}. \tag{19.10}$$

Here B_1 is the first Brillouin zone in the inverse lattice space.

As a result, we obtain the quadratic part of the phonon action functional S_{ph} in the following form:

$$S_{\mathrm{ph}}[\mathbf{P}_l, \mathbf{q}_l] = \beta^{-1} \sum_{\omega,\mathbf{p}\in B_1} [\omega\mathbf{p}_p(\omega)\mathbf{q}_{-p}(-\omega) - (2M)^{-1}\mathbf{p}_p(\omega)\mathbf{p}_{-p}(-\omega)$$
$$-\tfrac{1}{2}D_{ij}(\omega, \mathbf{p})(\mathbf{q}_p(\omega))_i(\mathbf{q}_{-p}(-\omega))_j]. \tag{19.11}$$

Here $D_{ij}(\omega, \mathbf{p})$ is the dynamical matrix

$$D_{ij}(\omega, \mathbf{p}) = M\omega_p^2\left\{\frac{p_i p_j}{p^2\varepsilon(\omega, \mathbf{p})} + \sum_{0\neq\mathbf{k}\in L^*}\left[\frac{(\mathbf{p}+\mathbf{k})_i(\mathbf{p}+\mathbf{k})_j}{(\mathbf{p}+\mathbf{k})^2\varepsilon(\omega, \mathbf{p}+\mathbf{k})} - \frac{k_i k_k}{k^2\varepsilon(\mathbf{k})}\right]\right\} \tag{19.12}$$

and

$$\omega_p^2 = \frac{4\pi e^2\rho}{M} \tag{19.13}$$

is a square of the ion-plasma frequency.

The last term in (19.5) is the electron–phonon interaction, which can be shown to be equal to

$$S_{\mathrm{el-ph}} = \beta^{-1}\sum_{\omega}\int d^3x e\delta\phi(\mathbf{x}, \mathbf{q}_l(\omega))\rho_0(\mathbf{x}, -\omega). \tag{19.14}$$

This is just the electrostatic energy of electrons near the Fermi surface in the electric potential due to the displacement of ions from their equilibrium positions.

Now we can obtain the effective interaction between electrons in the shell by integrating with respect to the phonon variables $\mathbf{p}_p(\omega), \mathbf{q}_p(\omega)$. First we rewrite $S_{\text{el-ph}}$ from (19.14) as follows:

$$S_{\text{el-ph}} = \beta^{-1} \sum_{\omega, \mathbf{p} \in B_1} (\mathbf{q}_{\mathbf{p}}(\omega), \mathbf{B}(-\mathbf{p}, -\omega)), \tag{19.15}$$

where

$$B(-\mathbf{p}, -\omega) = -\mathrm{i}N^{1/2}\frac{4\pi e^2}{\beta V}\sum_{\mathbf{k}\in L^*}\frac{\mathbf{p}+\mathbf{k}}{(\mathbf{p}+\mathbf{k})^2\varepsilon(\omega, \mathbf{p}+\mathbf{k})}\rho_0(-\mathbf{p}-\mathbf{k}, -\omega). \tag{19.16}$$

We denote by $M\omega_\lambda^2(\omega, \mathbf{p})$ the eigenvalues and by $\mathbf{e}_\lambda(\omega, \mathbf{p})$ the eigenvectors of the dynamical matrix. We have

$$D_{\alpha\beta}(\omega, \mathbf{p})\mathbf{e}_{\lambda\beta}(\omega, \mathbf{p}) = M\omega_\lambda^2(\omega, \mathbf{p})\mathbf{e}_{\lambda\alpha}(\omega, \mathbf{p}). \tag{19.17}$$

One can see that the eigenvalues obey the following sum rule:

$$\sum_{\lambda=1}^{3} \omega_\lambda^2(\omega, \mathbf{p}) = \omega_{\mathrm{p}}^2\varepsilon^{-1}(\omega, \mathbf{p}) + \omega_{\mathrm{p}}^2 \sum_{0\neq \mathbf{k}\in L^*} (\varepsilon^{-1}(\omega, \mathbf{p}+\mathbf{k}) - \varepsilon^{-1}(\mathbf{k})). \tag{19.18}$$

We may write the result of the integration with respect to the phonon variables in the following form:

$$\int Dp Dq \exp(S_{\text{ph}} + S_{\text{el-ph}}) = \exp \tilde{S}_{\text{el-el}}, \tag{19.19}$$

where

$$\tilde{S}_{\text{el-el}} = \frac{1}{2\beta}\sum_{\omega, \mathbf{p}\in B_1}\sum_{\alpha,\beta}(D(\omega, \mathbf{p}) - M(\mathrm{i}\omega)^2)^{-1}_{\alpha\beta}B_\alpha(\mathbf{p}, \omega)B_\beta(-\mathbf{p}, -\omega), \tag{19.20}$$

$$B_\alpha(\mathbf{p}, \omega)B_\beta(-\mathbf{p}, -\omega)$$
$$= M\omega_p^2\frac{4\pi e^2}{V}\sum_{\mathbf{k},\mathbf{k}'\in L^*}\frac{(\mathbf{p}+\mathbf{k})_\alpha(\mathbf{p}+\mathbf{k})_\beta\rho_0(\mathbf{p}+\mathbf{k}, \omega)\rho_0(-\mathbf{p}-\mathbf{k}', -\omega)}{(\mathbf{p}+\mathbf{k})^2(\mathbf{p}+\mathbf{k}')^2\varepsilon(\omega, \mathbf{p}+\mathbf{k})\varepsilon(\omega, \mathbf{p}+\mathbf{k}')} \tag{19.21}$$

Using the eigenvector basis of the dynamical matrix we get

$$(D(\omega, \mathbf{p}) - M(\mathrm{i}\omega)^2)^{-1}_{\alpha\beta} = M^{-1}(\omega_\alpha^2(\omega, \mathbf{p}) - (\mathrm{i}\omega)^2)^{-1}\delta_{\alpha\beta}. \tag{19.22}$$

Expression (19.20) describes the interaction between electrons near the Fermi surface due to the phonon exchange. A Coulomb repulsive term must be added to (19.20) in order to obtain the full effective interaction of electrons. This term has the following form:

$$S^{\mathrm{c}}_{\text{el-el}} = -\frac{1}{2\beta}\sum_{\omega}\int\frac{\mathrm{d}^3p}{(2\pi)^3}\frac{4\pi e^2}{p^2\varepsilon(\omega, \mathbf{p})}\rho_0(\mathbf{p}, \omega)\rho_0(-\mathbf{p}, -\omega). \tag{19.23}$$

Here we have replaced the sum over $\mathbf{p}, \mathbf{k}$ by the integral according to the rule:

$$V^{-1} \sum_{\mathbf{p} \in B_1, \mathbf{k} \in L^*} \to (2\pi)^{-3} \int d^3 p.$$

It is possible to simplify (19.20) a little by putting $\mathbf{k} = \mathbf{k}'$ in (19.21). It means that we neglect 'Umklapp' processes. This is justified if we are interested only in the interaction of electrons of opposite momenta near the Fermi surface. Assuming this we may rewrite (19.20) as

$$\tilde{S}_{\text{el-el}} = \frac{1}{2\beta V} \sum_{\omega, \alpha} \int \frac{d^3 p}{(2\pi)^3} \frac{4\pi e^2 \omega_p^2}{\omega_\alpha^2(\omega, \mathbf{p}) - (i\omega)^2} \frac{(\mathbf{p}, \mathbf{e}_\alpha(\omega, \mathbf{p}))^2}{p^4 \varepsilon^2(\omega, \mathbf{p})} \rho_0(\mathbf{p}, \omega) \rho_0(-\mathbf{p}, -\omega). \tag{19.24}$$

So we come down to the following effective electron–electron interaction:

$$S_{\text{eff el-el}} = -\frac{1}{2\beta} \sum_{\omega} \int \frac{d^3 p}{(2\pi)^3} \frac{4\pi e^2}{p^2 \varepsilon(\omega, \mathbf{p})} \left[1 - \sum_{\alpha} \frac{\omega_p^2 (\mathbf{p}, \mathbf{e}_\alpha)^2}{(\omega_\alpha^2 - (i\omega)^2) p^2 \varepsilon(\omega, p)} \right]$$
$$\times \rho_0(\mathbf{p}, \omega) \rho_0(-\mathbf{p}, -\omega). \tag{19.25}$$

We write down the effective electron–electron potential in the static limit

$$V_{\text{eff}}(\mathbf{p}) = V^c(\mathbf{p}) + V^{\text{ph}}(\mathbf{p}) = \frac{4\pi e^2}{p^2 \varepsilon(\mathbf{p})} \left(1 - \sum_{\alpha} \frac{\omega_p^2 p_\alpha^2}{\omega_\alpha^2(\mathbf{p}) p^2 \varepsilon(\mathbf{p})} \right). \tag{19.26}$$

In the large density limit $\varepsilon(\mathbf{p}) \to 1$ and (19.18) goes into

$$\sum_{\alpha=1}^{3} \omega_\alpha^2(\mathbf{p}) = \omega_{\mathbf{p}}^2. \tag{19.27}$$

So we may write (19.26) as

$$V_{\text{eff}}(\mathbf{p}) = \frac{4\pi e^2}{p^2} \left(1 - \sum_{\alpha} \frac{\omega_p^2 p_\alpha^2}{\omega_\alpha^2(\mathbf{p}) p^2} \right). \tag{19.28}$$

According to (19.27) we have $(\omega_p^2 / \omega_\alpha^2(\mathbf{p})) > 1$, and

$$\sum_{\alpha} \frac{\omega_p^2 p_\alpha^2}{\omega_\alpha^2(\mathbf{p}) p^2} > \sum_{\alpha} \frac{p_\alpha^2}{p^2} = 1. \tag{19.29}$$

This implies that the effective electron–electron potential is negative, and that electrons attract each other near the Fermi surface. The attractive interaction is the reason for the system to become a superconductor at sufficiently low temperatures.

20

Crystal structure of a dense electron–ion system

The functional integral approach to the theory of crystals allows us to find the energetically most preferable periodic lattice for a dense electron–ion system. For other methods in this problem see Abrikosov (1960, 1961) and Brovman, Kagan & Holas (1971, 1972). The approach developed below was suggested by Anisimov & Popov (1985). In order to solve this problem it is sufficient to compare the expression for the pressure $p = -\Omega/V$ for different crystalline structures and to choose a structure with the maximal value of p. In the case of large pressures (and densities) we have

$$p = p_0(\beta V)^{-1} S_{\text{eff}}^{(0)} - (2\beta V)^{-1} \ln \det (A/A_0) - (2V)^{-1} \sum_{\alpha, \mathbf{p} \in B_1} \omega_\alpha(\mathbf{p}) + \cdots \quad (20.1)$$

Here p_0 is the pressure of an ideal Fermi gas of electrons with chemical potential λ_e, $S_{\text{eff}}^{(0)}$ is the 'condensate' value of the effective action functional

$$\begin{aligned} S_{\text{eff}}[\mathbf{p}_{\mathbf{l},\alpha}, \mathbf{q}_{\mathbf{l},\alpha}, \varphi(\mathbf{x},\tau)] &= -(8\pi)^{-1} \int d\tau d^3x (\nabla\varphi(\mathbf{x},\tau))^2 \\ &+ 2 \ln \det [\partial_\tau + (2m)^{-1}\nabla^2 + \lambda_e + ie\varphi(x,\tau)]/[\partial_\tau + (2m)^{-1}\nabla^2 + \lambda_e] \\ &+ \int_0^\beta d\tau \sum_{\mathbf{l},\alpha} [\mathbf{p}_{\mathbf{l},\alpha}(\tau)\partial_\tau \mathbf{q}_{\mathbf{l},\alpha}(\tau) - (2M)^{-1}\mathbf{p}_{\mathbf{l},\alpha}^2(\tau) - ie\varphi(\mathbf{q}_{\mathbf{l},\alpha}(\tau),\tau)] \end{aligned} \quad (20.2)$$

for the external values

$$\mathbf{p}_{\mathbf{l},\alpha} = 0, \quad \mathbf{q}_{\mathbf{l},\alpha} = \mathbf{l} + \boldsymbol{\alpha}, \quad \mathbf{l} \in L, \quad \varphi(\mathbf{x},\tau) = -i\phi(\mathbf{x}). \quad (20.3)$$

Here $\mathbf{l} \in L$ are points of crystal lattice; the vectors $\boldsymbol{\alpha}$ show the location for ions inside the elementary cell; $\mathbf{p}_{\mathbf{l},\alpha}$, $\mathbf{q}_{\mathbf{l},\alpha}$ are the momentum and coordinate of an ion vibrating around a point $\mathbf{l} + \boldsymbol{\alpha}$; $\varphi(\mathbf{x},\tau)$ is the electric potential field.

The third term on the right-hand side of (20.1) is a result of functional integration with respect to the electric potential field in the vicinity of its condensate value $-i\phi(\mathbf{x})$. The operator A is defined by the formula for its matrix element

$$A(\mathbf{x},\omega,\mathbf{y},\omega') = \delta_{\omega\omega'}(-\nabla^2\delta(\mathbf{x}-\mathbf{y}) - 4\pi e^2 \Pi(\mathbf{x},\mathbf{y},\omega)). \quad (20.4)$$

Here

$$\Pi(\mathbf{x},\mathbf{y},\omega)=\beta^{-1}\sum_{\omega_1,s}G_s(\mathbf{x},\mathbf{y},\omega_1|e\phi)G_s(\mathbf{y},\mathbf{x},\omega+\omega_1|e\phi) \qquad (20.5)$$

is the so-called 'polarization operator', G_s is the Green function for an electron in the field $e\phi(\mathbf{x})$.

The last term $-(2V)^{-1}\sum_{\alpha,\mathbf{p}\in B_1}\omega_\alpha(\mathbf{p})$ in (20.1) is a contribution to the pressure from the zero vibrations energy of the crystal lattice; $\omega_\alpha(\mathbf{p})$ is the frequency of a phonon with momentum $\mathbf{p}$ and polarization α (B_1 is the first Brillouin zone in the inverse lattice space).

Let us consider each term on the right-hand side of (20.1). The expression for p_0 is well known:

$$p_0=2T(2\pi)^{-3}\int d^3k\ln(1+e^{-\beta\varepsilon(\mathbf{k})}), \qquad (20.6)$$

where i.e. expression for p_0 is (20.6) and that for $\varepsilon(\mathbf{k})$ is (20.7)

$$\varepsilon(\mathbf{k})=\frac{k^2}{2m}-\lambda_e. \qquad (20.7)$$

It can be calculated explicitly at low temperatures ($T\ll\varepsilon_F=\lambda_e=k_F^2/2m$) as

$$p_0=2(2\pi)^{-3}\int_{k^2<2m\lambda_e}d^3k\left(\lambda_e-\frac{k^2}{2m}\right)=\frac{2(2m)^{3/2}\lambda_e^{5/2}}{15\pi^2}=\frac{k_F^5}{15m\pi^2}. \qquad (20.8)$$

At low temperatures we have $\rho=k_F^3/3\pi^2$, and

$$p_0=\frac{A}{m}\rho^{5/3},\quad A=\tfrac{1}{5}(3\pi^2)^{2/3}. \qquad (20.9)$$

We see that p_0 is proportional to $\rho^{5/3}$. We will show that both the second and the third terms on the right-hand side of (20.1) are proportional to $\rho^{4/3}$. Hence they are small compared with p_0 at large pressures and densities.

The addends to the terms entering in (20.1) can be expressed via an infinite sum of diagrams of the modified perturbation theory. One can show that these contributions are small in comparison with the terms contained entering in (20.1) at large pressures and densities.

Now let us calculate the third term in (20.1) by expanding ln det into power series in Π and retaining only the first term of these series:

$$-\frac{1}{2\beta V}\ln\det A/A_0\approx\frac{e^2}{2\beta V}\iint\frac{d^3x\,d^3y}{|\mathbf{x}-\mathbf{y}|}\sum_\omega\Pi(\mathbf{x},\mathbf{y},\omega). \qquad (20.10)$$

We have

$$\beta^{-1}\sum_\omega\Pi(\mathbf{x},\mathbf{y},\omega)$$

$$=\beta^{-2}\sum_{s,\omega,\omega_1}G_s(\mathbf{x},\mathbf{y},\omega_1|e\phi)G_s(\mathbf{y},\mathbf{x},\omega+\omega_1|e\phi)$$

$$= \sum_s \left(\beta^{-1} \sum_{\omega_1} G_s(\mathbf{x}, \mathbf{y}, \omega_1 | e\phi) \right) \left(\beta^{-1} \sum_{\omega_1} G_s(\mathbf{y}, \mathbf{x}, \omega_2 | e\phi) \right). \tag{20.11}$$

Now we may use the following formula for G_s:

$$G_s(\mathbf{x}, \mathbf{y}, \omega | e\phi) = \sum_\alpha \frac{\psi_{\alpha s}(\mathbf{x}) \bar{\psi}_{\alpha s}(\mathbf{y})}{i\omega - E_\alpha + \lambda_e}, \tag{20.12}$$

where $\sum_\alpha$ is a sum over the complete system of eigenfunctions of the operator $-(2m)^{-1}\nabla^2 + e\phi(\mathbf{x})$ with eigenvalues E_α. We can perform summation over ω_1, ω_2 in (20.11) as follows:

$$\begin{aligned} \beta^{-1} \sum_\omega G_s(\mathbf{x}, \mathbf{y}, \omega | e\phi) &= \lim_{\varepsilon \to +0} \beta^{-1} \sum_\omega e^{i\omega\varepsilon} G_s(\mathbf{x}, \mathbf{y}, \omega | e\phi) \\ &= \lim_{\varepsilon \to +0} \sum_\alpha \psi_{\alpha s}(\mathbf{x}) \bar{\psi}_{\alpha s}(\mathbf{y}) \beta^{-1} \sum_\omega \frac{e^{i\omega\varepsilon}}{i\omega - E_\alpha + \lambda_e} \\ &= \sum_\alpha n_\alpha \psi_{\alpha s}(\mathbf{x}) \bar{\psi}_{\alpha s}(\mathbf{y}), \end{aligned} \tag{20.13}$$

where $n_\alpha = (\exp \beta(E_\alpha - \lambda_e) + 1)^{-1}$. At low temperatures ($T \to 0$) we obtain

$$\beta^{-1} \sum_\omega G_s(\mathbf{x}, \mathbf{y}, \omega | e\phi) = \sum_{E_\alpha < \lambda_e} \psi_{\alpha s}(\mathbf{x}) \bar{\psi}_{\alpha s}(\mathbf{y}) \tag{20.14}$$

and (20.10) goes into

$$-\frac{1}{2\beta V} \ln \det A/A_0 \approx \frac{e^2}{2V} \iint \frac{d^3x d^3y}{|\mathbf{x} - \mathbf{y}|} \sum_{\substack{s, \alpha, \gamma \\ E_\alpha, E_\gamma < \lambda_e}} \psi_{\alpha s}(\mathbf{x}) \bar{\psi}_{\alpha s}(\mathbf{y}) \psi_{\gamma s}(\mathbf{y}) \bar{\psi}_{\gamma s}(\mathbf{x}). \tag{20.15}$$

This is nothing but the expression for the electron exchange energy divided by $-V$.

In the high density limit the eigenfunctions $\psi_{\alpha s}(\mathbf{x})$ turn into the eigenfunctions for an ideal gas

$$\psi_{\alpha s}(\mathbf{x}) = V^{-1/2} e^{i\mathbf{k}\mathbf{x}}. \tag{20.16}$$

Substituting (20.16) and

$$|\mathbf{x} - \mathbf{y}|^{-1} = 4\pi V^{-1} \sum_{k \neq 0} k^{-2} e^{i(\mathbf{k}, \mathbf{x} - \mathbf{y})} \tag{20.17}$$

into (20.15), we find

$$\begin{aligned} -\frac{1}{2\beta V} \ln \det A/A_0 &= 4\pi e^2 V^{-2} \sum_{\substack{k_1^2, k_2^2 < k_F^2 \\ \mathbf{k}_1 \neq \mathbf{k}_2}} |\mathbf{k}_1 - \mathbf{k}_2|^{-2} \\ &= \frac{4\pi e^2}{(2\pi)^6} \iint_{k_1^2, k_2^2 < k_F^2} \frac{d^3k_1 d^3k_2}{|\mathbf{k}_1 - \mathbf{k}_2|^2} = \frac{e k_F^4}{4\pi^3} = \frac{B}{2} e^2 \rho^{4/3}, \end{aligned} \tag{20.18}$$

where

$$B = \frac{3}{2}\left(\frac{3}{\pi}\right)^{1/3} = 1.47711753\ldots . \tag{20.19}$$

One can see that the exchange term is proportional to $e^2\rho^{4/3}$ and does not depend on the structure of the crystal lattice. If we calculate (20.15) with greater accuracy, we obtain the higher-order corrections to (20.18) which depend on the crystal lattice type.

The most interesting term in (20.1) is $S^{(0)}_{\text{eff}}/\beta V$. This term turns out to be of the order $e^2\rho^{4/3}$ as well as the exchange term, but the coefficient of $e^2\rho^{4/3}$ does depend on the type of the lattice. This allows us to choose the most preferable lattice in the high density limit.

In order to find $S^{0}_{\text{eff}}/\beta V$ let us substitute (20.3) into (20.2). Expanding ln det in (20.2) into power series in ϕ and retaining the second-order terms, we have

$$S^{(0)}_{\text{eff}}/\beta V = V^{-1}\left\{(8\pi)^{-1}\int(\nabla\phi)^2\,\mathrm{d}^3x - e\int \mathrm{d}^3x\phi(\mathbf{x})\left(\sum_{\mathbf{l},\boldsymbol{\alpha}}\delta(\mathbf{x}-\mathbf{l}-\boldsymbol{\alpha})-\rho\right)\right.$$

$$\left. - e^2\int \mathrm{d}\tau\,\mathrm{d}^3x\,\mathrm{d}^3y\,\phi(\mathbf{x})\phi(\mathbf{y})G_0(\mathbf{x}-\mathbf{y},\tau)G_0(\mathbf{y}-\mathbf{x},-\tau)\right\}. \tag{20.20}$$

Here

$$\rho = 2G_0(\mathbf{x}-\mathbf{y},\tau)|_{\mathbf{x}=\mathbf{y},\tau=+0} = \frac{N}{V} \tag{20.21}$$

is the electron gas density, G_0 is the Green function of a free electron. If we use the self-consistent field equation for $\phi(\mathbf{x})$:

$$\nabla^2\phi(\mathbf{x}) + 8\pi e^2\int \mathrm{d}\tau\,\mathrm{d}^3y\,G_0(\mathbf{x}-\mathbf{y},\tau)G_0(\mathbf{y}-\mathbf{x},-\tau)\phi(\mathbf{y})$$

$$= 4\pi e\left\{\rho - \sum_{\mathbf{l}\in\mathbf{L},\boldsymbol{\alpha}}\delta(\mathbf{x}-\mathbf{l}-\boldsymbol{\alpha})\right\} \tag{20.22}$$

we may rewrite (20.20) in the form

$$S^{(0)}_{\text{eff}}/\beta V = -\frac{e}{2V}\int \mathrm{d}^3x\phi(\mathbf{x})\left(\sum_{\mathbf{l},\boldsymbol{\alpha}}\delta(\mathbf{x}-\mathbf{l}-\boldsymbol{\alpha})-\rho\right). \tag{20.23}$$

Equation (20.22) has the following solution:

$$\phi(\mathbf{x}) = \frac{4\pi e\rho}{s}\sum_{\substack{0\neq\mathbf{k}\in L^*\\ \boldsymbol{\alpha}}}\frac{e^{i(\mathbf{k},\mathbf{x}-\mathbf{y})}}{k^2\varepsilon(\mathbf{k})}. \tag{20.24}$$

Here L^* is the inverse lattice, s is the number of ions in the elementary cell, $\varepsilon(\mathbf{k})$ is the dielectric function of an electron gas in RPA. Let us substitute (20.24) into (20.23). The term $-\rho$ in (20.23) gives a zero contribution because the integral of $e^{i(\mathbf{k},\mathbf{x}-\boldsymbol{\alpha})}$ over the elementary cell vanishes for $\mathbf{k} \neq 0$. So we obtain

$$-\frac{e}{2V}\sum_{\mathbf{l},\boldsymbol{\alpha}} \phi(\mathbf{l}+\boldsymbol{\alpha}) = -\frac{N_c}{V}\frac{2\pi e^2 \rho}{s} \sum_{\substack{\boldsymbol{\alpha},\boldsymbol{\beta} \\ 0 \neq \mathbf{k} \in L^*}} \frac{e^{i(\mathbf{K},\boldsymbol{\alpha}-\boldsymbol{\beta})}}{k^2 \varepsilon(\mathbf{k})}, \tag{20.25}$$

where N_c is the number of cells in the crystal. This expression is divergent because of terms with $\boldsymbol{\alpha} = \boldsymbol{\beta}$. That is why we have to regularize this result replacing $\boldsymbol{\alpha}$ in (20.25) by $\boldsymbol{\alpha} + \boldsymbol{\varepsilon}$, adding $Ne^2/2|\boldsymbol{\varepsilon}|\,V$ (the energy per volume of $N = N_c s$ metallic spheres of radius $|\boldsymbol{\varepsilon}|$) and then taking the limit $|\boldsymbol{\varepsilon}| \to 0$. This regularization corresponds to the transition from the model of point-like ions to the model of small spheres of radius $|\boldsymbol{\varepsilon}|$. So we obtain

$$(S^{(0)}_{\text{eff}}/\beta V)_{\text{reg}} = -\lim_{|\boldsymbol{\varepsilon}|\to 0}\left(\frac{N_c}{V}\frac{2\pi e^2 \rho}{s} \sum_{\substack{\boldsymbol{\alpha},\boldsymbol{\beta} \\ 0 \neq \mathbf{k} \in L^*}} \frac{e^{i(\mathbf{k},\boldsymbol{\varepsilon}+\boldsymbol{\alpha}-\boldsymbol{\beta})}}{k^2 \varepsilon(\mathbf{k})} - \frac{Ne^2}{2V|\boldsymbol{\varepsilon}|}\right)$$

$$= -\frac{\rho e^2}{2}\lim_{|\boldsymbol{\varepsilon}|\to 0}\left(\frac{4\pi}{sV} \sum_{\substack{\boldsymbol{\alpha},\boldsymbol{\beta} \\ 0 \neq \mathbf{k} \in L^*}} \frac{e^{i(\mathbf{k},\boldsymbol{\varepsilon}+\boldsymbol{\alpha}-\boldsymbol{\beta})}}{k^2 \varepsilon(\mathbf{k})} - \frac{1}{|\boldsymbol{\varepsilon}|}\right). \tag{20.26}$$

Let us consider a high density limit $\varepsilon(\mathbf{k}) \to 1$. We will show how to calculate (20.26) for lattices consisting of one or more cubic sublattices. Let a be the length of the edge; $a^3 = v$, the cell volume. Substituting

$$\mathbf{k} = \frac{2\pi}{a}\mathbf{n}, \quad \boldsymbol{\alpha} = a\boldsymbol{\gamma} \quad \boldsymbol{\beta} = a\boldsymbol{\delta}, \quad \boldsymbol{\varepsilon} = a\boldsymbol{\xi} \tag{20.27}$$

into (20.27) and putting $\varepsilon(\mathbf{k}) = 1$, we have

$$(S^{(0)}_{\text{eff}}/\beta V)_{\text{reg}} = -\frac{e^2}{2}\rho^{4/3}\lim_{|\boldsymbol{\xi}|\to 0} s^{-1/3}\left(\sum_{\substack{\mathbf{n}\neq 0 \\ \boldsymbol{\gamma},\boldsymbol{\delta}}} \frac{\exp 2\pi i(\mathbf{n}, \boldsymbol{\xi}+\boldsymbol{\gamma}-\boldsymbol{\delta})}{s\pi n^2} - |\boldsymbol{\xi}|^{-1}\right). \tag{20.28}$$

The easiest way to evaluate (20.28) is to exploit the Evald method based on the Poisson summation formula. The starting point of the method is the identity

$$(\pi n^2)^{-1} = \int_0^\infty d\zeta\, e^{-\zeta\pi n^2} = \int_0^\eta + \int_\eta^\infty = \int_0^\eta d\zeta\, e^{-\zeta\pi n^2} + (\pi n^2)^{-1} e^{-\eta\pi n^2}. \tag{20.29}$$

Here η is an arbitrary real positive number (the so-called Evald parameter). We may rewrite (20.28) as

$$-\frac{e^2}{2}\rho^{4/3}\lim_{|\xi|\to 0} s^{-1/3}\Big\{s^{-1/3}\sum_{\mathbf{n}\neq 0,\gamma,\delta}(\pi n^2)^{-1}\exp(2\pi i(\mathbf{n},\xi+\gamma-\delta)-\eta\pi n^2)$$

$$-s^{-1}\sum_{\mathbf{n}\neq 0,\gamma,\delta}\int_0^\eta d\zeta\exp(2\pi i(\mathbf{n},\xi+\gamma-\delta)-\zeta\pi n^2)-|\xi|^{-1}\Big\}. \qquad (20.30)$$

We may add a term corresponding to $\mathbf{n}=0$ to the second sum in this formula. Now we can modify the expression

$$I=\sum_{\mathbf{n}}\exp(-\zeta\pi n^2+2\pi i(\mathbf{n},\boldsymbol{\kappa})), \qquad (20.31)$$

where $\boldsymbol{\kappa}=\xi+\gamma-\delta$ by using the Poisson transformation,

$$I=\int d^3x\exp(-\zeta\pi x^2+2\pi i(\mathbf{x},\boldsymbol{\kappa}))\Big(\sum_{\mathbf{n}}\delta(\mathbf{x}-\mathbf{n})\Big)$$

$$=\int d^3x\exp(-\zeta\pi x^2+2\pi i(\mathbf{x},\boldsymbol{\kappa}))\Big(\sum_{\mathbf{m}}\exp 2\pi i(\mathbf{m},\mathbf{x})\Big)$$

$$=\sum_{\mathbf{m}}\zeta^{-3/2}\exp\Big(-\frac{\pi}{\zeta}(\boldsymbol{\kappa}+\mathbf{m})^2\Big). \qquad (20.32)$$

Substituting this into (20.30), we obtain

$$-\frac{e^2}{2}\rho^{4/3}\lim_{|\xi|\to 0} s^{-1/3}\Big\{s^{-1}\sum_{\mathbf{n}\neq 0,\gamma,\delta}(\pi n^2)^{-1}\exp(2\pi i(\boldsymbol{\kappa},\mathbf{n})-\eta\pi n^2)$$

$$+s^{-1}\sum_{\mathbf{m},\gamma,\delta}\int_0^\eta d\zeta\zeta^{-3/2}\exp\Big(-\frac{\pi}{\zeta}(\mathbf{m}+\boldsymbol{\kappa})^2\Big)-\eta s-|\xi|^{-1}\Big\}. \qquad (20.33)$$

The terms in the second sum corresponding to $\mathbf{m}=0$, $\gamma=\delta$, give the contribution

$$\int_0^\eta d\zeta\zeta^{-3/2}\exp\Big(-\frac{\pi}{\zeta}|\xi|^2\Big)$$

$$=\int_0^\infty d\zeta\zeta^{-3/2}\exp\Big(-\frac{\pi}{\zeta}|\xi|^2\Big)-\int_\eta^\infty d\zeta\zeta^{-3/2}\exp\Big(-\frac{\pi}{\zeta}|\xi|^2\Big). \qquad (20.34)$$

The first integral on the right-hand side of (20.34) is equal to $|\xi|^{-1}$. This term cancels with $-|\xi|^{-1}$ in (20.33). Now we may put $\xi=0$ in all other terms. In particular, the integral

$$\int_{\eta}^{\infty} d\zeta \zeta^{-3/2} \exp\left(-\frac{\pi}{\zeta}|\xi|^2\right)$$

turns out to be equal to $\int_{\eta}^{\infty} d\zeta\zeta^{-3/2} = 2\eta^{-1/2}$.

So we come down to the following formula:

$$(S_{\text{eff}}^{(0)}/\beta V)_{\text{reg}} = -\frac{e^2}{2}\rho^{3/3}s^{-1/3}\left\{s^{-1}\sum_{\mathbf{n}\neq 0,\gamma,\delta}(\pi n^2)^{-1}\exp(-\eta\pi n^2 + 2\pi i(\mathbf{n}, \gamma - \delta))\right.$$

$$\left. + s^{-1}\sum_{\mathbf{m}+\gamma-\delta\neq 0}|\mathbf{m}+\gamma-\delta|^{-1}\operatorname{erf}\left(\left(\frac{\pi}{\eta}\right)^{1/2}|\mathbf{m}+\gamma-\delta|\right) - \eta s - 2\eta^{-1/2}\right\}. \tag{20.35}$$

Here

$$\operatorname{erf} x = \left(\frac{2}{\pi}\right)^{1/2}\int_{x}^{\infty} e^{-t^2/2}\, dt. \tag{20.36}$$

This formula can be applied to the numerical calculation of $(S_{\text{eff}}^{(0)}/\beta V)_{\text{reg}}$ for lattices consisting of one or several cubic sublattices – the simple cubic (SC) lattice, the cubic body centered (CBC) lattice and the cubic face centred (CFC) lattice. The number of ions in the elementary cell and all possible values of γ-vectors for the three lattices are

$$\begin{aligned}
&\text{SC} && s = 1, && \gamma = 0, \\
&\text{CBC} && s = 2, && \gamma_1 = 0, \quad \gamma_2 = (\tfrac{1}{2}, \tfrac{1}{2}, \tfrac{1}{2}), \\
&\text{CFC} && s = 4, && \gamma_1 = 0, \quad \gamma_2 = (0, \tfrac{1}{2}, \tfrac{1}{2}), \\
&&&&& \gamma_3 = (\tfrac{1}{2}, 0, \tfrac{1}{2}), \quad \gamma_4 = (\tfrac{1}{2}, \tfrac{1}{2}, 0).
\end{aligned} \tag{20.37}$$

Here we regard CBC and CFC lattices as consisting of several cubic sublattices, so that there are two ions in one elementary cell in CBC and four ions in CFC. It is well known that each of these lattices can be regarded as a lattice with only ion in their elementary cell; this is not, however, a cubic lattice. For instance, the CBC lattice is generated by the vectors $(1,0,0)$, $(0,1,0)$, $(\frac{1}{2},\frac{1}{2},\frac{1}{2})$, the CFC lattice is generated by three vectors $\gamma_2, \gamma_3, \gamma_4$ in (20.37). Nevertheless, the method used here seems more useful for numerical calculations.

If the Evald parameter η is close to unity, the series in (20.35) converges rapidly. The results of calculations are

$$\begin{aligned}(S^{(0)}_{\text{eff}}/\beta V)_{\text{reg}} &= \tfrac{1}{2}e^2\rho^{4/3}2.83729748\ldots \quad \text{SC}\\ &= \tfrac{1}{2}e^2\rho^{4/3}2.88846150\ldots \quad \text{CBC}\\ &= \tfrac{1}{2}e^2\rho^{4/3}2.88828212\ldots \quad \text{CFC}.\end{aligned} \tag{20.38}$$

We see that the CBC lattice gives the largest contribution to the pressure p, and contribution of the CFC lattice is very close to that of the CBC lattice. The simple cubic lattice SC gives a smaller contribution to p than either the CBC or CFC lattices. This is why the SC lattice is less preferable than the CBC and CFC lattices. If we wish to consider an arbitrary crystalline lattice, it is useful to go over to dimensionless variables:

$$\mathbf{k} = 2\pi\mathbf{g}v^{-1/3}, \quad \boldsymbol{\alpha} = v^{1/3}\boldsymbol{\gamma}, \quad \boldsymbol{\beta} = v^{1/3}\boldsymbol{\delta}, \quad \boldsymbol{\varepsilon} = v^{1/3}\boldsymbol{\xi}. \tag{20.39}$$

Here $v = ([\mathbf{r}_1, \mathbf{r}_2]\mathbf{r}_3)$, $\mathbf{r}_i (i = 1, 2, 3)$ are basis vectors of the lattice L; $\boldsymbol{\gamma}, \boldsymbol{\delta}$ are dimensionless vectors corresponding to ions in an elementary cell; $\mathbf{k}$ are the inverse lattice vectors. It is not difficult to generalize (20.28) to an arbitrary lattice:

$$\begin{aligned}(S^{(0)}_{\text{eff}}/\beta V)_{\text{reg}} &= \frac{e^2}{2}\rho^{4/3}\lim_{|\xi|\to 0} s^{-1/3}\\ &\times\left\{s^{-1}\sum_{\mathbf{g}\neq 0,\gamma,\delta}(\pi g^2)^{-1}\exp(2\pi\mathrm{i}(\mathbf{g}, \boldsymbol{\xi} + \boldsymbol{\gamma} - \boldsymbol{\delta})) - |\boldsymbol{\xi}|^{-1}\right\}.\end{aligned} \tag{20.40}$$

By means of Evald's method this formula can be transformed to the following form:

$$\begin{aligned}(S^{(0)}_{\text{eff}}/\beta V)_{\text{reg}} &= \frac{e^2}{2}\rho^{4/3}s^{-1/3}\\ &\times\left\{s^{-1}\sum_{\mathbf{g}\neq 0,\gamma,\delta}(\pi g^2)^{-1}\exp(-\eta\pi g^2 + 2\pi\mathrm{i}(\mathbf{g}, \boldsymbol{\gamma} - \boldsymbol{\delta}))\right.\\ &+ s^{-1}\sum_{\mathbf{b}+\gamma-\delta\neq 0}|\mathbf{b} + \boldsymbol{\gamma} - \boldsymbol{\delta}|^{-1}\operatorname{erf}\left(\left(\frac{\pi}{\eta}\right)^{1/2}|\mathbf{b} + \boldsymbol{\gamma} - \boldsymbol{\delta}|\right)\\ &\left. - \eta s - 2\eta^{-1/2}\right\}.\end{aligned} \tag{20.41}$$

Here $\mathbf{g}$ and $\mathbf{b}$ are the vectors of mutually inverse lattices, and we have

$$\begin{aligned}\mathbf{g} = n_1\mathbf{g}_1 + n_2\mathbf{g}_2 + n_3\mathbf{g}_3, \quad \mathbf{b} = m_1\mathbf{b}_1 + m_2\mathbf{b}_2 + m_3\mathbf{b}_3\\ (\mathbf{g}_i\,\mathbf{b}_j) = \delta_{ij}, \quad i, j = 1, 2, 3.\end{aligned} \tag{20.42}$$

If η is close to unity, both series in (20.41) are rapidly convergent. Expression (20.41) was used for the calculation of $(S^{(0)}_{\text{eff}}/\beta V)_{\text{reg}}$ for certain lattices. Besides cubic structures the simple hexagonal lattice (SH) and the closed packed hexagonal structure (HCP) were considered. The ratio c/a is the length of the shortest vertical lattice vector; ε, the length of the shortest lattice vector in the horizontal plane, was chosen on the condition that $(S^{(0)}_{\text{eff}}/\beta V)_{\text{reg}}$ be maximal. The simple tetragonal (ST) lattices and the simple rhombohedric lattices (SR) were studied for different values of $c/a, c/b$ and α, where $\mathbf{a}, \mathbf{b}$ are the lattice vectors in the horizontal plane, α is the angle between $\mathbf{a}$ and $\mathbf{b}$, and the maximum value of $(S^{(0)}_{\text{eff}}/\beta V)_{\text{reg}}$ was found as the maximum of a function of three variables $c/a, c/b$ and α.

The results of the calculation of the values $C = 2(S^{(0)}_{\text{eff}}/\beta V)_{\text{reg}}/e^2\rho^{4/3}$ are as follows:

SC	2.83729747948...		
CBC	2.88846150305...		
CFC	2.88828211902...		
SH	2.86070968169...	$c/a = 0.9283$	(20.43)
HPC	2.88816875048...	$c/a = 1.6357$	
diamond type	2.69339902207...		
ST	2.83729747948...		
SR	2.85919037116...		

One can see that the CBC structure is the preferable one. The CFC and HCP structures (at $c/a = 1.635$ close to an 'ideal' value $(c/a)_{\text{id}} = (8/3)^{1/2}$) have values close to that of the CBC. The SH structure at $c/a \approx 0.93$ has the next value of C.

The SC structure has the maximal value of $(S^{(0)}_{\text{eff}})_{\text{reg}}$ among tetragonal lattices. The lattice with $c/a = c/b = 0.9$, $\alpha \approx 57°$ is the most preferable one among rhombic lattices.

The calculations show that the structures with main vectors which have very different lengths are not preferable.

So, if we take into account the first three terms in (20.1), we obtain the following formula:

$$p = p_0 + \tfrac{1}{2}e^2\rho^{4/3}(B + C) + \cdots. \tag{20.44}$$

Here p_0 is an ideal gas pressure; the constant B in (20.44) is universal (see (20.19)); C depends on the type of lattice. Corrections to (20.44) may be responsible for the so-called structure phase transitions. The possibility of such transitions is very probably due to the very small difference in C values for the CBC and CFC lattices (1.8×10^{-4}).

One of the corrections to (20.44) is the last term in (20.1). This is the so-called zero vibrations energy: the inequality

$$\left(V^{-1}\sum_{\mathbf{p}\in B_1,\alpha}\omega_\alpha(\mathbf{p})\right)^2 \leqslant \left(V^{-1}\sum_{\mathbf{p}\in B_1,\alpha}1\right)\left(V^{-1}\sum_{\mathbf{p}\in B_1,\alpha}\omega_\alpha^2(\mathbf{p})\right) \tag{20.45}$$

and the sum rule

$$\sum_\alpha \omega_\alpha^2(\mathbf{p}) = \omega_p^2\left[\varepsilon^{-1}(\mathbf{p}) + \sum_{0\neq\mathbf{k}\in L^*}(\varepsilon^{-1}(\mathbf{p}+\mathbf{k}) - \varepsilon^{-1}(\mathbf{k}))\right]. \tag{20.46}$$

($\omega_p = (4\pi e^2\rho M^{-1})^{1/2}$ is the ion-plasmon frequency) show that at large pressures

$$V^{-1}\sum_{\mathbf{p}\in B_1,\alpha}\omega_\alpha(\mathbf{p}) \sim eM^{-1/2}\rho^{3/2}. \tag{20.47}$$

We see that the contribution from the zero vibrations energy contains terms of higher order in the density $\rho^{3/2}$ as compared with $\rho^{4/3}$ in (20.44). But, in fact, the presence of the small parameter $M^{-1/2}$ ensures that the contribution of the zero vibrations energy is small in comparison with terms entering in (20.44).

There exists another correction

$$\frac{2\pi e^2\rho}{Vs}\sum_{\substack{0\neq\mathbf{k}\in L^*\\ \boldsymbol{\alpha},\boldsymbol{\beta}}} k^{-2}(1-\varepsilon^{-1}(\mathbf{k}))\exp i(\mathbf{k},\boldsymbol{\alpha}-\boldsymbol{\beta}) \tag{20.48}$$

which takes into account the fact that $\varepsilon(\mathbf{k})$ differs from unity.

Still another correction to (20.44) must arise if we take into account the fact that the electron eigenfunctions $\psi_{\alpha s}, \psi_{\gamma s}$ in (20.15) differ from the eigenfunctions of an ideal gas (20.16).

The above-mentioned corrections may give rise to the possible structure phase transitions between different lattices.

21

Quantum crystals

The functional integral approach to the theory of crystals outlined in previous sections is founded on the idea of a nontrivial average of the electric potential field $\langle \varphi(\mathbf{x}, \tau) \rangle \neq 0$. This approach is immediately applicable only to crystals consisting of particles with Coulomb interaction. In this section we extend this approach to systems with a short-range pair interaction between particles. Such systems may develop into a crystalline state, if the potential has an attractive part.

Let us consider a Fermi system with the action functional

$$S = \sum_{p,s} \left(i\omega - \frac{q^2}{2m} + \lambda \right) a_s^*(p) a_s(p) - (2\beta V)^{-1} \sum_p U(p) \rho(p) \rho(-p). \quad (21.1)$$

Here $p = \{\mathbf{q}, \omega\}$; λ is the chemical potential;

$$\rho(p) = \sum_{p_1, s} a_s^*(p + p_1) a_s(p_1); \quad (21.2)$$

$U(p) = U(q)$ is the Fourier transform of the pair potential function. We suppose $U(q)$ to have the following form (see Fig. 21.1):

$$\begin{aligned} U(q) &= U_+(q) \geqslant 0 \quad \text{if } q \in [0, k_1] \\ &= U_-(q) \leqslant 0 \quad \text{if } q \in [k_1, k_2] \\ &= 0 \qquad\qquad\quad\ \text{if } q > k_2. \end{aligned} \quad (21.3)$$

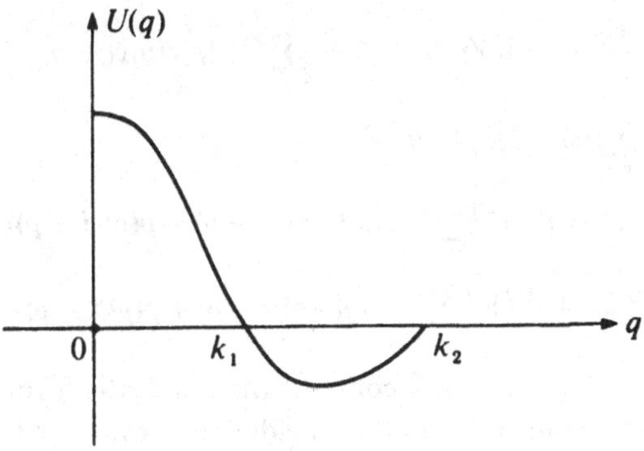

Fig. 21.1

Another approach to this problem was developed by Kirjnitz & Nepomniachij (1970).

Consider the functional integral

$$\int \mathrm{D}a^*\mathrm{D}a \exp S \tag{21.4}$$

which is proportional to the partition sum of the system in question. Let us insert the Gaussian integral with respect to some auxiliary Bose field $b(p)$ and $c(p)$ into the functional integral (21.4).

$$\mathrm{D}b\mathrm{D}c \exp\left(-\tfrac{1}{2}\sum_p U_+(p)b(p)b(-p) + \tfrac{1}{2}\sum_p U_-(p)c(p)c(-p) \right). \tag{21.5}$$

Here $b(p)$, $c(p)$ are the Fourier coefficients of the fields $b(\mathbf{x}, \tau)$, $c(\mathbf{x}, \tau)$ defined by the following formulae:

$$b(\mathbf{x}, \tau) = (\beta V)^{-1/2} \sum_p e^{ipx} b(p),$$

$$c(\mathbf{x}, \tau) = (\beta V)^{-1/2} \sum_p e^{ipx} c(p), \tag{21.6}$$

$px = \omega\tau + \mathbf{qx}$. The signs before $U_\pm(p)$ are chosen in such a way that the Gaussian integral (21.5) makes sense. In (21.5) and (21.6) summation over q is performed according to the definition of $U_\pm(q)$ in (21.3), i.e. $q \in [0, k_1]$ for $b(p)$ and $q \in [k_1, k_2]$ for $c(p)$.

Now we can make the shift transformation of the auxiliary Bose fields $b(p)$, $c(p)$

$$b(p) \to b(p) + i(\beta V)^{-1/2} \rho(p)$$

$$c(p) \to c(p) + (\beta V)^{-1/2} \rho(p). \tag{21.7}$$

After this transformation we will have in place of (21.1) the following action:

$$\begin{aligned} \tilde{S} = & -\frac{1}{2}\sum_p U_+(p)b(p)b(-p) + \frac{1}{2}\sum_p U_-(p)c(p)c(-p) \\ & + \sum_{p,s} (i\omega - \xi) a_s^*(p) a_s(p) \\ & + (2(\beta V)^{1/2})^{-1} \sum_p U_-(p)(c(p)\rho(-p) + \rho(p)c(-p)) \\ & - i(2(\beta V)^{1/2})^{-1} \sum_p U_+(p)(b(p)\rho(-p) + \rho(p)b(-p)), \end{aligned} \tag{21.8}$$

where $\xi = q^2/2m - \lambda$. This action $\tilde{S}$ contains the quadratic form of Fermi fields $a_s(p)$, $a_s^*(p)$, so the integral over these fields can be evaluated explicitly. Integrating over this Fermi field and regularizing the determinant, we obtain the following effective action functional:

$$S_{\text{eff}} = -\frac{1}{2}\sum_p U_+(p)b(p)b(-p) + \frac{1}{2}\sum_p U_-(p)c(p)c(-p)$$
$$+ 2\ln\det M(b,c)/M(0,0). \tag{21.9}$$

Here the factor 2 before ln det is due to the summation over the spin index $s = \pm$ and $M(b, c)$ is defined by the formula

$$M_{p_1p_2}(b,c) = (i\omega - \xi)\delta_{p_1p_2} - i(\beta V)^{-1/2}U_+(p_1 - p_2)b(p_1 - p_2)$$
$$+ (\beta V)^{-1/2}U_-(p_1 - p_2)c(p_1 - p_2). \tag{21.10}$$

The effective action functional (21.9) contains information on many properties of a model system, especially on the spectrum of Bose collective excitations.

We begin the investigation of S_{eff} with the uniform state of the system. In this state there is a Bose condensate of the field $b(\mathbf{x}, \tau)$, and we suppose that the main contribution to the functional integral with respect to the Bose fields $b(\mathbf{x}, \tau)$, $c(\mathbf{x}, \tau)$, comes from the vicinity of fields

$$b(\mathbf{x},\tau) = -ib_0, \quad c(\mathbf{x},\tau) = 0, \tag{21.11}$$

or in the p representation

$$b(p) = b_0(p) = -i(\beta V)^{1/2}b_0\delta_{p0}, \quad c(p) = 0. \tag{21.12}$$

Here b_0 is some parameter, which can be found approximately by replacing the field $b(p)$, $c(p)$ in S_{eff} (21.3) by their condensate values (21.12). So we have

$$S^{(0)}_{\text{eff}} = \tfrac{1}{2}\beta V U_+(0)b_0^2 + 2\ln\det[(i\omega - \xi - U_+(0))/(i\omega - \xi)\delta_{p_1p_2}]. \tag{21.13}$$

We can find b_0 from $\partial S^{(0)}_{\text{eff}}/\partial b_0 = 0$, or

$$b_0 = \frac{2}{\beta V}\sum_{\mathbf{q},\omega}(i\omega - \xi - U_+(0)b_0)^{-1} = \lim_{\varepsilon\to+0}\frac{2}{\beta V}\sum_{\mathbf{q},\omega}\frac{e^{i\omega\varepsilon}}{i\omega - \xi - U_+(0)b_0}. \tag{21.14}$$

After the summation over Fermi frequencies $\omega = (2n+1)\pi/\beta$ and taking the limit $\varepsilon \to +0$, we get the equation

$$b_0 = \frac{2}{V}\sum_{\mathbf{q}}\left(\exp\beta\left(\frac{q^2}{2m} - \lambda + U_+(0)b_0\right) + 1\right)^{-1}. \tag{21.15}$$

It is not difficult to see that there exists a solution b_0 of (21.15). Putting

$$\lambda = \frac{k_F^2}{2m} + U_+(0)b_0 \tag{21.16}$$

we obtain

$$b_0 = \frac{N}{V}, \quad N = 2\sum_{\mathbf{q}}\left(\exp\frac{\beta}{2m}(q^2 - k_F^2) + 1\right)^{-1}. \tag{21.17}$$

Here N is the number of Fermi particles.

Now we shall derive the dispersion law of low-energy collective excitations of the system ($q/k_F \ll 1$). In order to do it, we extract from S_{eff} the quadratic from Q of fluctuations of fields $b(p), c(p)$ near their condensate values (21.12). This quadratic form may be written down as

$$-\frac{1}{2}\sum_p b(p)b(-p)U_+(p)\left(1 - \frac{2U_+(p)}{\beta V}\sum_{\mathbf{q}_1,\omega_1}(i\omega_1 - \varepsilon_1)^{-1}(i\omega + i\omega_1 - \varepsilon_2)^{-1}\right)$$
$$+\frac{1}{2}\sum_p c(p)c(-p)U_-(p)\left(1 - \frac{2U_-(p)}{\beta V}\sum_{\mathbf{q}_1,\omega_1}(i\omega_1 - \varepsilon_1)^{-1}(i\omega + i\omega_1 - \varepsilon_2)^{-1}\right). \tag{21.18}$$

Here $\varepsilon_1 = \varepsilon(\mathbf{q}_1)$, $\varepsilon_2 = \varepsilon(\mathbf{q}_2)$, $\varepsilon(\mathbf{q}) = (q^2 - k_F^2)/2m$.

According to (21.18) we can write down a Green function of both $b(p)$ and $c(p)$ fields in the first approximation in the following form:

$$D(p) = |U(q)|^{-1}(1 - U(q)\Pi(p))^{-1}, \tag{21.19}$$

where

$$\Pi(p) = [\text{loop}] = \frac{2}{\beta V}\sum_{p_1}(i\omega_1 - \varepsilon_1)^{-1}(i\omega + i\omega_1 - \varepsilon_2)^{-1} \tag{21.20}$$

is a polarization operator of the Fermi system. We have to take into account that the three-momenta of the b-field belong to the interval $[0, k_1]$, and the three-momenta of the field c belong to $[k_1, k_2]$. After the summation over the Fermi frequencies in (21.20) we get

$$\beta^{-1}\sum_{\omega_1}(i\omega_1 - \varepsilon_1)^{-1}(i\omega + i\omega_1 - \varepsilon_2)^{-1} = (i\omega - \varepsilon_2 + \varepsilon_1)^{-1}(n_1 - n_2), \tag{21.21}$$

where $n_i = n(q_i)$ are the Fermi–Dirac partition function.

For small momenta $q(q/k_F \ll 1)$ we have

$$n_1 - n_2 \approx -\frac{\partial n}{\partial q_1} q\cos\theta, \tag{21.22}$$

where θ is the angle between the vectors $\mathbf{q}$ and $\mathbf{q}_1$.

Equating the denominator of Green's function (21.19) to zero, we obtain the following dispersion equation:

$$1+\frac{2U(q)}{V}\sum_{q_1}\frac{q\cos\theta}{i\omega-\varepsilon_2+\varepsilon_1}\frac{\partial n}{\partial q_1}=0. \tag{21.23}$$

Let us change $i\omega$ to E and replace the sum by the integral. Integrating with respect to angle variables, we get

$$\frac{m^2}{(2\pi)^2}\int_0^{\infty}\frac{E}{q}\ln\frac{E-\dfrac{qq_1}{m}}{E+\dfrac{qq_1}{m}}\frac{\partial n}{\partial q_1}dq_1=\frac{1}{2U(q)}-\frac{m^2}{2\pi^2}\int_0^{\infty}\frac{\partial n}{\partial q_1}q_1\,dq_1. \tag{21.24}$$

This equation implies the existence of phonon excitations in the system. Let us demonstrate this in the case $T=0$. In this case $\partial n/\partial q=-\delta(q-k_F)$, and we can rewrite (21.24) as follows:

$$\left(\frac{m}{2\pi}\right)^2\frac{E}{q}\ln\frac{E+\dfrac{k_F q}{m}}{E-\dfrac{k_F q}{m}}=\frac{mk_F}{2\pi^2}+\frac{1}{2U(q)}. \tag{21.25}$$

Putting $E=cq$, we recover for small q ($U(q)\approx U(0)$), the well-known Landau dispersion equation for the zero sound velocity c:

$$\frac{mc}{2k_F}\ln\frac{c+k_F/m}{c-k_F/m}-1=\frac{\pi^2}{mk_F U(0)}. \tag{21.26}$$

Denoting mc/k_F by x, we may rewrite (21.26) as

$$f(x)=\frac{x}{2}\ln\frac{x-1}{x-1}-1=\frac{\pi^2}{mk_F U(0)}. \tag{21.27}$$

Hence $f(1+0)=+\infty$, $f(+\infty)=0$, and $f(x)$ decreases on the interval $(1,\infty)$. So (21.27) has a single root for $x\in(1,\infty)$. It means that the zero sound velocity is larger than the velocity on the Fermi surface $c_F=k_F/m$.

Now we will discuss the problem of stability of the uniform state. The stability condition can be written in the following form:

$$D^{-1}(0,q)>0 \tag{21.28}$$

or

$$1-U(q)\Pi(0,q)>0. \tag{21.29}$$

The uniform state becomes unstable if for some $q = k_0$

$$U_-(k_0)\Pi(0, k_0) = 1 \tag{21.30}$$

or

$$\frac{2U_-(k_0)}{V}\sum_{q_1}\frac{n(\mathbf{q}_1) - n(\mathbf{q}_1 + \mathbf{k}_0)}{\varepsilon(\mathbf{q}_1) - \varepsilon(\mathbf{q}_1 + \mathbf{k}_0)} = 1. \tag{21.31}$$

If $T = 0$, we have $n(q) = \theta(k_F - q)$, and (21.31) has the following form:

$$-\frac{U_-(k_0)mk_F}{2\pi^2}\left[\frac{k_F}{k_0}\left(1 - \frac{k_0^2}{4k_F^2}\right)\ln\left|\frac{2k_F + k_0}{2k_F - k_0}\right| + 1\right] = 1. \tag{21.32}$$

Here the function in $[\ldots]$ in (21.32) is non-negative for all $k_0 \in [0, \infty)$. This function takes values in the interval $(0, 2)$ (it goes to 2 if $k_0 \to 0$ and vanishes if $k_0 \to \infty$). So in order for the system to lose stability the interaction has to be sufficiently strong (and negative). Hence we must have

$$-\frac{U_-(k_0)N}{\varepsilon_F V} \sim 1. \tag{21.33}$$

At low temperatures ($T/\varepsilon_F \ll 1$) we can write (21.31) as

$$-\frac{U_-(k_0)mk_F}{2\pi^2}\left[\frac{k_F}{k_0}\left(1 - \frac{k_0^2}{4k_F^2}\right)\ln\left|\frac{2k_F + k_0}{2k_F - k_0}\right| + 1 - \frac{mT^2}{3k_F^2(k_F^2 - k_0^2/4)}\right] = 1. \tag{21.34}$$

We see that if the temperature increases, the stability violation takes place at a higher density than for $T = 0$ (if $k_F > k_0/2$). What happens if the stability conditions (21.28) and (21.29) are not fulfilled? It is natural to expect that the system will go to the crystalline state. We will describe such a state by the condensate functions

$$b = -ib_0, \quad c = \sum_{0 \neq \mathbf{k} \in L^*} c_{\mathbf{k}} e^{i\mathbf{k}\mathbf{x}}, \tag{21.35}$$

where L^* is a lattice inverse to L. Vectors $\mathbf{k} \in L^*$ in (21.35) must obey the same condition $|\mathbf{k}| \in [k_1, k_2]$, as the three-momenta of field $c(p)$. This is why the sum in (21.35) contains a finite number of terms. We will suppose for the sake of simplicity that this sum contains only those vectors of L^* which are nearest neighbours to $\mathbf{k} = 0$, and that all of them have equal length.

It is natural to take the $c_{\mathbf{k}}$ also to be equal to each other. So we may rewrite (21.35) in p-representation as follows:

$$b(p) = -ib_0(\beta V)^{1/2}\delta_{p0}, \quad c(p) = c_0(\beta V)^{1/2}\delta_{\omega,0}\sum_{\mathbf{k}_0}\delta_{\mathbf{q},\mathbf{k}_0}, \tag{21.36}$$

where $\sum_{\mathbf{k}_0}$ is the sum over nearest neighbours to the zero lattice point $\mathbf{k} = 0$ in L^*.

One can consider b_0 and c_0 in (21.36) as variational parameters, which have to be found from the maximum conditions for the condensate value of S_{eff}:

$$S^{(0)}_{\text{eff}} = \tfrac{1}{2}\beta V U_+(0) b_0^2 + \tfrac{1}{2}\beta V U_-(0) c_0^2 \left(\sum_{\mathbf{k}_0} 1 \right)$$
$$+ 2 \ln \det \frac{(i\omega - \xi - b_0 U_+(0))\delta_{p_1 p_2} + U_-(k_0) c_0 \delta_{\omega_1 \omega_2} \sum_{\mathbf{k}_0} \delta_{\mathbf{q}_1 - \mathbf{q}_2, \mathbf{k}_0}}{(i\omega - \xi)\delta_{p_1 p_2}} \tag{21.37}$$

These maximum conditions are as follows:

$$b_0 = \frac{2}{\beta V} \operatorname{Tr} \left[(i\omega - \xi - b_0 U_+(0))\delta_{p_1 p_2} + c_0 U_-(k_0) \delta_{\omega_1 \omega_2} \sum_{\mathbf{k}_0} \delta_{\mathbf{q}_1 - \mathbf{q}_2, \mathbf{k}_0} \right]^{-1},$$
$$c_0 \left(\sum_{\mathbf{k}_0} 1 \right) = \frac{2}{\beta V} \operatorname{Tr} \left(\delta_{\omega_1 \omega_2} \sum_{\mathbf{k}_0} \delta_{\mathbf{q}_1 - \mathbf{q}_2, \mathbf{k}_0} \right)$$
$$\cdot \left[(i\omega - \xi - b_0 U_+(0))\delta_{p_1 p_2} + c_0 U_-(k_0) \delta_{\omega_1 \omega_2} \sum_{\mathbf{k}_0} \delta_{\mathbf{q}_1 - \mathbf{q}_2, \mathbf{k}_0} \right]^{-1}. \tag{21.38}$$

The nonuniform crystal state corresponds to a nonzero value of c_0. If we suppose the state to be weakly nonuniform, c_0 is small, and we may expand the right-hand side of (21.38) in c_0. We then come to the following approximate equation which replaces the first equation in (21.38):

$$b_0 = \frac{2}{\beta V} \sum_{\mathbf{q},\omega} (i\omega - \xi - b_0 U_+(0))^{-1} + \frac{2 n c_0^2 U_-^2(k_0)}{\beta V}$$
$$\cdot \sum_{\mathbf{q},\omega} (i\omega - \xi - b_0 U_+(0))^{-1} (i\omega - \xi(\mathbf{q} + \mathbf{k}_0) - b_0 U_+(0))^{-2}, \tag{21.39}$$

where

$$n = \sum_{\mathbf{k}_0} 1 \tag{21.40}$$

is the number of nearest neighbours in L^*.

The first term on the right-hand side of (21.39) was calculated above (see (21.14) and (21.15)). It is equal to

$$\rho = \frac{2}{V} \sum_{\mathbf{q}} \left(\exp \frac{\beta}{2m}(q^2 - k_F^2) + 1 \right)^{-1}. \tag{21.41}$$

The second term on the right-hand side of (21.35) is equal to

$$(\beta V)^{-1}\sum_{\mathbf{q},\omega}(i\omega-\varepsilon_1)^{-1}(i\omega-\varepsilon_2)^{-2}=V^{-1}\sum_{\mathbf{q}}\left[\frac{n_1-n_2}{(\varepsilon_1-\varepsilon_2)^2}-\frac{\beta n_2(1-n_2)}{\varepsilon_2-\varepsilon_1}\right], \tag{21.42}$$

where $n_1=n(\mathbf{q}), n_2=n(\mathbf{q}+\mathbf{k}_0)$ are the Fermi–Dirac distribution functions. The term $(n_1-n_2)/(\varepsilon_1-\varepsilon_2)^2$ has no contribution to (21.42) due to its antisymmetry under the permutations $\varepsilon_1\rightleftarrows\varepsilon_2, n_1\rightleftarrows n_2$. If we make such permutations in $\beta n_2(1-n_2)/(\varepsilon_1-\varepsilon_2)$ as well, we obtain

$$(\beta V)^{-1}\sum_{\mathbf{q},\omega}(i\omega-\varepsilon_1)^{-1}(i\omega-\varepsilon_2)^{-2}=(2V)^{-1}\sum_{\mathbf{q}}\frac{\beta[n_2(1-n_2)-n_1(1-n_1)]}{\varepsilon_1-\varepsilon_2}. \tag{21.43}$$

It is not difficult to calculate (21.43) at low temperatures ($T\to 0$). In this case the main contribution to (21.43) comes from the regions with $\mathbf{q}$ or $\mathbf{q}+\mathbf{k}_0$ close to Fermi surface. So we obtain

$$(2V)^{-1}\sum_{\mathbf{q}}\frac{\beta[n_2(1-n_2)-n_1(1-n_1)]}{\varepsilon_1-\varepsilon_2}\approx\frac{mk_F}{(2\pi)^3}\int\frac{d\Omega d\xi\dfrac{\beta e^{\beta\xi}}{(e^{\beta\xi}+1)^2}}{\dfrac{(\mathbf{k}_F+\mathbf{k}_0)^2}{2m}-\dfrac{k_F^2}{2m}}$$

$$=\frac{m^2k_F}{4\pi^3}\int\frac{d\Omega}{k_0^2+2(\mathbf{k}_F\mathbf{k}_0)}=\frac{m^2k_F}{2\pi^2}\int_{-1}^{1}\frac{dx}{2k_Fk_0x+k_0^2} \tag{21.44}$$

and we may rewrite (21.39) as follows:

$$b_0=\frac{2}{V}\sum_{\mathbf{q}}n(\mathbf{q})+nc_0^2U_-^2(k_0)V^{-1}\sum_{\mathbf{q}}\frac{\beta[n_2(1-n_2)-n_1(1-n_1)]}{\varepsilon_1-\varepsilon_2}$$

$$=\rho+\frac{nc_0^2U_-^2(k_0)m^2}{2\pi^2k_0}\ln\left|\frac{2k_F+k_0}{2k_F-k_0}\right|. \tag{21.45}$$

Now let us consider the second equation in (21.38). Using the same approximation as for the first equation, we can rewrite it as

$$nc_0=\frac{2nc_0U_-(k_0)}{\beta V}\sum_{\mathbf{q},\omega}(i\omega-\varepsilon_1)^{-1}(i\omega-\varepsilon_2)^{-1}$$

$$-\frac{2c_0^3U_-^2(k_0)}{\beta V}\sum_{\mathbf{k}_{01},\mathbf{k}_{02},\mathbf{k}_{03}}\sum_{\mathbf{q},\omega}(i\omega-\varepsilon_1)^{-1}(i\omega-\varepsilon_2)^{-1}(i\omega-\varepsilon_3)^{-1}. \tag{21.46}$$

In the first term on the right-hand side of (21.46) we have $\varepsilon_1 = \varepsilon(\mathbf{q})$, $\varepsilon_2 = \varepsilon(\mathbf{q}+\mathbf{k}_0)$, and in the second one $\varepsilon_i = \varepsilon(\mathbf{q}_i)$, $i = 1, 2, 3$, where $\mathbf{q}_1, \mathbf{q}_2, \mathbf{q}_3$ are vectors obeying the following relations:

$$\mathbf{q}_1 - \mathbf{q}_2 = \mathbf{k}_{01}, \quad \mathbf{q}_2 - \mathbf{q}_3 = \mathbf{k}_{02}, \quad \mathbf{q}_3 - \mathbf{q}_1 = \mathbf{k}_{03}. \tag{21.47}$$

Equation (21.47) implies

$$\mathbf{k}_{01} + \mathbf{k}_{02} + \mathbf{k}_{03} = 0. \tag{21.48}$$

Let us note that this equality cannot be valid for an arbitrary inverse lattice L^*. For instance, among three lattices with cubic symmetry (SC, CBC, CFC) (21.48) can hold only if L is the CBC lattice (L^* is CFC).

The first term on the right-hand side of (21.46) is equal to

$$\frac{2nc_0U_-(k_0)}{V}\sum_{\mathbf{q}}\frac{n_1 - n_2}{\varepsilon_1 - \varepsilon_2}$$
$$\approx -\frac{nc_0U_-(k_0)mk_F}{\pi^2}\left[\frac{k_F}{k_0}\left(1 - \frac{k_0^2}{4k_F^2}\right)\ln\left|\frac{2k_F + k_0}{2k_F - k_0}\right| + 1 - \frac{m^2T^2}{k_F^2(k_F^2 - k_0^2/4)}\right]. \tag{21.49}$$

The second term on the right-hand side of (21.46) can be written down as follows:

$$-\frac{2c_0^2U_-^2(k_0)nn'}{V}\sum_{\mathbf{q}}\left[\frac{n_1}{(\varepsilon_1-\varepsilon_2)(\varepsilon_1-\varepsilon_3)} + \frac{n_2}{(\varepsilon_2-\varepsilon_3)(\varepsilon_2-\varepsilon_1)}\right.$$
$$\left. + \frac{n_3}{(\varepsilon_3-\varepsilon_1)(\varepsilon_3-\varepsilon_2)}\right] = -\frac{6c_0^2U_-^2(k_0)nn'}{V}\sum_{\mathbf{q}}\frac{n_1}{(\varepsilon_1-\varepsilon_2)(\varepsilon_1-\varepsilon_3)} \tag{21.50}$$

Hence n is the number of nearest neighbours in L^*; n', for a given $\mathbf{k}_{01}$, is the number of vectors $\mathbf{k}_{02}$ such that $\mathbf{k}_{01} + \mathbf{k}_{02} = -\mathbf{k}_{03}$. For example, $n = 12$, $n' = 4$ for a CBC lattice.

It is possible to calculate $\sum_{\mathbf{q}}$ in (21.50) using the Feynman procedure:

$$\frac{1}{V}\sum_{\mathbf{q}}\frac{n_1}{(\varepsilon_1-\varepsilon_2)(\varepsilon_1-\varepsilon_3)} = \frac{4m^2}{(2\pi)^3}\int_{q^2<k_F^2}\frac{d^3q}{(k_0^2 + 2(\mathbf{k}_{01},\mathbf{q}))(k_0^2 - 2(\mathbf{k}_{02},\mathbf{q}))}$$
$$= \frac{m^2}{2\pi^3}\int_0^1 d\alpha\int_{q^2<k_F^2}\frac{d^3q}{[k_0^2 + 2(\mathbf{k}_{01}\alpha - \mathbf{k}_{02}(1-\alpha),\mathbf{q})]^2}$$

$$= \frac{m^2}{6\pi^2 k_0}\left[k_0 \ln\left|\frac{2k_F + k_0}{2k_F - k_0}\right| - \tfrac{4}{3}(3k_F^2 - k_0^2)^{1/2} \right.$$

$$\left. \cdot\left(\frac{\pi}{2} - \arctan\frac{(3k_F^2 - k_0^2)^{1/2}}{k_F}\right)\right] \quad \text{if} \quad 0 < k_0 < k_F 3^{1/2},$$

$$= \frac{m^2}{6\pi^2 k_0}\left[k_0 \ln\left|\frac{2k_F + k_0}{2k_F - k_0}\right| \right.$$

$$\left. + \tfrac{2}{3}(k_0^2 - 3k_F^2)^{1/2} \ln\left|\frac{(k_0^2 - 3k_F^2)^{1/2} + k_F}{(k_0^2 - 3k_F^2)^{1/2} - k_F}\right|\right]$$

$$\text{if} \quad k_0 > k_F 3^{1/2}. \tag{21.51}$$

This is an increasing function of k_0, if $k_0 \in [0, 2k_F]$. This function is negative for $k_0 \to 0$ and positive if $k_0 > \bar{k} = 1.05 k_F$.

Now we can rewrite (21.46) as

$$nc_0 = \frac{2nc_0 U_-(k_0)}{V}\sum_q \frac{n_1 - n_2}{\varepsilon_1 - \varepsilon_2} - \frac{6c_0^2 U_-^2(k_0) nn'}{V}\sum_q \frac{n_1}{(\varepsilon_1 - \varepsilon_2)(\varepsilon_1 - \varepsilon_3)}. \tag{21.52}$$

Its solution is

$$c_0 = \left(1 - \frac{2U_-(k_0)}{V}\sum_q \frac{n_1 - n_2}{\varepsilon_1 - \varepsilon_2}\right) \Big/ \left(-\frac{6U_-^2(k_0)n'}{V}\sum_q \frac{n_1}{(\varepsilon_1 - \varepsilon_2)(\varepsilon_1 - \varepsilon_3)}\right). \tag{21.53}$$

Here the numerator becomes negative if the stability conditions (21.28) and (21.29) are violated. The denominator of the right-hand side of (21.53) becomes negative when k_0 exceeds 1.05 k_F and we have a positive solution (21.53) in this case. If we know c_0 we may then find b_0 using (21.45). One can check that (21.45) and (21.53) correspond to the maximum of $S_{\text{eff}}^{(0)}$. These results give evidence in favour of stability of the weak-nonuniform state with the CBC lattice. Nevertheless, it can be shown that this state is instable under small fluctuations.

We can show this by considering the second variation form

$$Q = -\frac{1}{2}\sum_p U_+(p)b(p)b(-p) + \frac{1}{2}\sum_p U_-(p)c(p)c(-p) - \operatorname{Tr}(UG)^2. \tag{21.54}$$

Here

$$(G^{-1})_{p_1 p_2} = (i\omega - \xi - b_0 U_+(0))\delta_{p_1 p_2} + U_-(k_0)c_0 \delta_{\omega_1 \omega_2}\sum_{k_0} \delta_{q_1 - q_2, k_0}, \tag{21.55}$$

$$U_{p_1p_2} = i(\beta V)^{-1/2}U_+(p_1 - p_2)b(p_1 - p_2) - (\beta V)^{-1/2}U_-(p_1 - p_2)c(p_1 - p_2).$$

We have

$$Q = Q_1(b, b) + Q_2(c, c) + Q_{12}(b, c), \tag{21.56}$$

where Q_1 depends only on b-fields, Q_2 on c-fields; Q_{12} is linear in both b and c.

$$Q_1(b, b) = -\frac{1}{2}\sum_p b(p)b(-p)\left(U_+(p) - \frac{2U_+^2(p)}{\beta V}\sum_{p_1}(i\omega_1 - \varepsilon_1)^{-1}(i\omega + \varepsilon_2)^{-1}\right) + \text{terms of the order of } c_0^2. \tag{21.57}$$

Then we have

$$Q_{12}(b, c) \sim c_0. \tag{21.58}$$

It means that the form Q_{12} is proportional to a small parameter c_0. The form $Q_2(c, c)$ is also proportional to c_0. This implies that the 'link' between the b-mode and the c-mode is proportional to c_0^2, i.e. it is small compared to c_0. So in the first approximation we can neglect this link, i.e. we can omit $Q_{12}(b, c)$.

We shall be mostly interested in $Q_2(c, c)$, which can be written as

$$Q_2(c, c) = Q_{20}(c, c) + Q_{21}(c, c). \tag{21.59}$$

Here Q_{20} does not depend on c_0 and Q_{21} is linear in c_0. We neglect the terms proportional to c_0^2, c_0^3 and so on. The form Q_{20} is the same as for a uniform system, i.e.

$$Q_{20} = \frac{1}{2}\sum_p c(p)c(-p)\left(U_-(p) - \frac{2U_-^2(p)}{\beta V}\sum_{p_1}(i\omega_1 - \varepsilon_1)^{-1}(i\omega + i\omega_1 - \varepsilon_2)^{-1}\right). \tag{21.60}$$

Q_{21} can be expressed as

$$Q_{21}(c, c) = \sum_{k_0}\sum_{p+p'+p_0=0} U_-(p)U_-(p')c(p)c(p')\frac{2c_0U_-(k_0)}{\beta V} \sum_{\substack{p_1-p_2=p\\ p_2-p_3=p_0\\ p_3-p_1=p_0}} G(p_1)G(p_2)G(p_3). \tag{21.61}$$

Here $p_0 = (k_0, 0)$ is a four-vector with zero frequency component, and its momentum component is equal to one of the nearest neighbour vectors

$\mathbf{k}_0 \in L^*$. Now we can answer the question about the stability of the CBC lattice. Let us take in $Q_2 = Q_{20} + Q_{21}$ only the terms corresponding to $p = p_0 = (\mathbf{k}_0, 0)$. This is a sum of a finite number of terms

$$\frac{1}{2}\sum_{p_0} c(p_0)c(-p_0)U_-(k_0)\left(1 - \frac{2U_-(k_0)}{V}\sum_{q}\frac{n_1 - n_2}{\varepsilon_1 - \varepsilon_2}\right)$$

$$+ \sum_{p_{01}+p_{02}+p_{03}=0} c(p_{01})c(p_{02})\frac{6c_0 U_-^2(k_0)}{V}\sum_{q}\frac{n_1}{(\varepsilon_1 - \varepsilon_2)(\varepsilon_1 - \varepsilon_3)}. \quad (21.62)$$

Using (21.53) we can rewrite (21.62) as

$$\tfrac{1}{2}U_-(k_0)\left(1 - \frac{2U_-(k_0)}{V}\sum_{q}\frac{n_1 - n_2}{\varepsilon_1 - \varepsilon_2}\right)$$

$$\left[\sum_{p_0} c(p_0)c(-p_0) - \frac{2}{n'}\sum_{p_{01}+p_{02}+p_{03}=0} c(p_{01})c(p_{02})\right]. \quad (21.63)$$

The factor before [...] in (21.63) is positive in the region of nonstability of the uniform state. In order that the nonuniform state with CBC lattice is stable it is necessary for (21.63) to be negative. So the expression

$$\sum_{p_0} c(p_0)c(-p_0) - \frac{2}{n'}\sum_{p_{01}+p_{02}+p_{03}=0} c(p_{01})c(p_{02}) \quad (21.64)$$

must be a nonpositive quadratic form. This form will be negative if we put $c(p_0) = c$ for all $\mathbf{k}_0 \in L^*$. In this case we obtain

$$c^2\left(\sum_{p_0} 1 - \frac{2}{n'}\sum_{p_{01}+p_{02}+p_{03}=0} 1\right) = nc^2(1-2) = -nc^2 < 0. \quad (21.65)$$

Nevertheless, if we take $c(p_0) = c(-p_0) = c \neq 0$ for only one of six directions corresponding to vectors $\pm\mathbf{k}_0$ in the CBC lattice and $c(p_0) = 0$ for all other directions, the second sum in (21.64) vanishes, and the first one is equal to $2c^2 > 0$. This result shows that (21.64) is not a nonpositive form. Physically this means that the weak-nonuniform state with the CBC lattice is unstable. This is in agreement with the well-known Landau statement (Landau, 1937) on the impossibility of the second-order phase transition of a uniform phase to a nonuniform one with a lattice L for which there exist three nearest neighbour vectors in L^* satisfying $\mathbf{k}_{01} + \mathbf{k}_{02} + \mathbf{k}_{03} = 0$.

We see that the lattice associated with the weak-nonuniform state is such that $\mathbf{k}_{01} + \mathbf{k}_{02} + \mathbf{k}_{03} = 0$ is impossible in L^*. As for lattices with cubic symmetry, only SC (simple cubic) and CFC (cubic face centred) may be lattices of a weak-nonuniform state.

22
The theory of heavy atoms

In this section we will use functional integral methods in order to describe systems with a large but finite number of particles. We can take the many electron atoms as an example of such a system. It is also interesting to apply functional methods to the atomic nucleus.

The functional integral approach allows us to build up the ground state of the system considered and also to obtain information on the collective excitations. Below we dwell only on the theory of heavy atoms with many ($Z \gg 1$) electrons (see Bidzhelova & Popov, 1982).

We begin with the formula for the quotient of partition functions for nonideal and ideal Fermi systems.

$$\frac{Z}{Z_0} = \frac{\int e^{S} \mathrm{D}\bar{\psi}\mathrm{D}\psi}{\int e^{S_0} \mathrm{D}\bar{\psi}\mathrm{D}\psi} \tag{22.1}$$

is the formal quotient of two functional integrals over Fermi fields $\psi(\mathbf{x}, \tau)$, $\bar{\psi}(\mathbf{x}, \tau)$. Here S is the action functional of electrons in the Coulomb potential of nucleus:

$$S = \int_0^\beta \mathrm{d}\tau \int \mathrm{d}^3x \sum_s \Bigg(\bar{\psi}_s(\mathbf{x},\tau)\partial_\tau \psi_s(\mathbf{x},\tau) - (2m)^{-1}\nabla\bar{\psi}_s(\mathbf{x},\tau)\nabla\psi_s(\mathbf{x},\tau)$$
$$+ \left(\lambda + \frac{Ze^2}{r}\right)\bar{\psi}_s(\mathbf{x},\tau)\psi_s(\mathbf{x},\tau)\Bigg) - \frac{1}{2}\int_0^\beta \mathrm{d}\tau \int \mathrm{d}x\,\mathrm{d}y \frac{e^2}{|\mathbf{x}-\mathbf{x}|}\rho(\mathbf{x},\tau)\rho(\mathbf{y},\tau),$$

where

$$\rho(\mathbf{x},\tau) = \sum_s \bar{\psi}_s(\mathbf{x},\tau)\psi_s(\mathbf{x},\tau) \tag{22.3}$$

is the electron density, $\tau \in [0, \beta]$, $\beta = T^{-1}, s = \pm$ is the spin index.

As for S_0 it will be taken in the form

$$S_0 = \int_0^\beta d\tau \int d^3x \sum_s (\bar{\psi}_s(\mathbf{x},\tau)\partial_\tau \psi_s(\mathbf{x},\tau)$$
$$- (2m)^{-1}\nabla\bar{\psi}_s(\mathbf{x},\tau)\nabla\psi_s(\mathbf{x},\tau) - V(\mathbf{x})\bar{\psi}_s(\mathbf{x},\tau)\psi_s(\mathbf{x},\tau)), \qquad (22.4)$$

where $V(\mathbf{x})$ is some attractive potential which will be specified below.

Now we use once again the trick which was previously applied to other systems of particles with the Coulomb interaction. We introduce into the functional integrals in the numerator and the denominator of (22.1) the following Gaussian integral over the Bose field $\varphi(\mathbf{x},\tau)$ of electric potential:

$$\int D\varphi \exp\left[-(8\pi)^{-1}\int d\tau\, d^3x(\nabla\varphi(\mathbf{x},\tau))^2\right]. \qquad (22.5)$$

Then we perform the shift transformation

$$\varphi(\mathbf{x},\tau) \to \varphi(\mathbf{x},\tau) + ie\int \frac{\rho(\mathbf{y},\tau)\,d^3y}{|\mathbf{x}-\mathbf{y}|} \qquad (22.6)$$

which allows us to cancel the nonlocal term of the Coulomb interaction, and we come to the following functional:

$$S[\psi,\bar{\psi},\varphi] = \int d\tau d^3x\Big\{\sum_s [\bar{\psi}_s(\mathbf{x},\tau)\partial_\tau\psi_s(\mathbf{x},\tau) - (2m)^{-1}\nabla\bar{\psi}_s(\mathbf{x},\tau)\nabla\psi_s(\mathbf{x},\tau)$$
$$+ (\lambda + Ze^2r^{-1} - ie\varphi(\mathbf{x},\tau))\bar{\psi}_s(\mathbf{x},\tau)\psi_s(\mathbf{x},\tau)]$$
$$- (8\pi)^{-1}(\nabla\varphi(\mathbf{x},\tau))^2\Big\}. \qquad (22.7)$$

It depends on ψ, $\bar{\psi}$, φ and describes the system of Fermi particles interacting via the field of scalar electric potential. Performing the Gaussian functional integration over Fermi fields, we obtain the following expression for Z/Z_0:

$$\frac{Z}{Z_0} = \frac{\int D\varphi \exp S_{\text{eff}}[\varphi]}{\int D\varphi \exp S_0[\varphi]}, \qquad (22.8)$$

where

$$S_{\text{eff}}[\varphi] = -(8\pi)^{-1}\int d\tau d^3x(\nabla\varphi)^2 + 2\ln\det(M[\varphi]/M[V]),$$

$$S_0[\varphi] = (8\pi)^{-1}\int d\tau d^3x(\nabla\varphi)^2, \qquad (22.9)$$

$$M[\varphi] = \partial_\tau + (2m)^{-1}\nabla^2 + \lambda + Ze^2r^{-1} - ie\varphi(\mathbf{x},\tau),$$

$$M[V] = \partial_\tau + (2m)^{-1}\nabla^2 + V(\mathbf{x}). \qquad (22.10)$$

It is natural to suppose that the ground state of a heavy atom corresponds to some average electric potential $\varphi_0(\mathbf{x},\tau) = \varphi_0(\mathbf{x})$, which can be obtained from the equation

$$\delta S_{\text{eff}}[\varphi_0] = 0 \tag{22.11}$$

or

$$\nabla^2\varphi_0(\mathbf{x},\tau) + 4\pi i e[\partial_\tau + (2m)^{-1}\nabla^2 + \lambda + Ze^2r^{-1} - ie\varphi_0(\mathbf{x},\tau)]^{-1}_{\mathbf{x},\tau;\mathbf{x},\tau} = 0. \tag{22.12}$$

Substituting

$$\varphi_0(\mathbf{x},\tau) = -i\phi(\mathbf{x}) \tag{22.13}$$

we may write (22.12) as

$$\nabla^2\phi = -4\pi e\rho(\mathbf{x}|-\lambda + e\phi(\mathbf{x}) - Ze^2r^{-1}) \tag{22.14}$$

where $\rho(\mathbf{x}|V)$ has the meaning of electron density in the field V.

Let us suppose Z to be even. Suppose also that there exist exactly $Z/2$ negative eigenvalues of the Schrödinger equation

$$(-(2m)^{-1}\nabla^2 - \lambda + e\phi(\mathbf{x}) - Ze^2r^{-1})\psi = E\psi. \tag{22.15}$$

Now we choose the potential $V(\mathbf{x})$ to be equal to

$$V(\mathbf{x}) = -\lambda + e\phi(\mathbf{x}) - Zer^{-1}. \tag{22.16}$$

If we make the shift transformation

$$\varphi(\mathbf{x},\tau) \to \varphi(\mathbf{x},\tau) - i\phi(\mathbf{x}) \tag{22.17}$$

we can write down $S_{\text{eff}}[\varphi]$ in the form

$$S_{\text{eff}}[\varphi] = (\beta/8\pi)\int(\nabla\phi)^2 d^3x - (8\pi)^{-1}\int d\tau\, d^3x(\nabla\varphi(\mathbf{x},\tau))^2$$

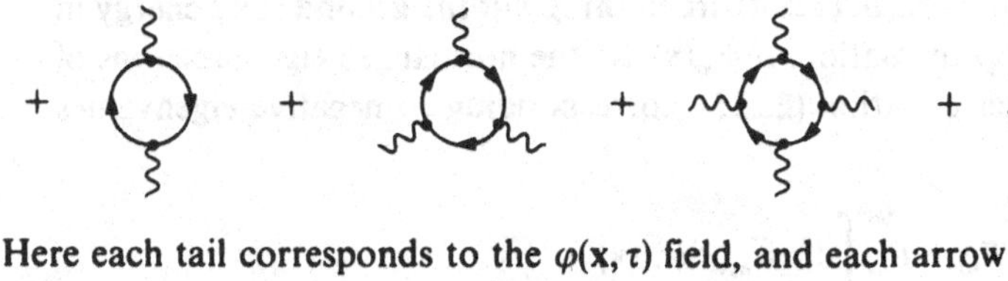

(22.18)

Here each tail corresponds to the $\varphi(\mathbf{x},\tau)$ field, and each arrow line $\longrightarrow$ corresponds to the Green function of an electron in the field $V(\mathbf{x})$. This Green function may be expressed as

$$G(\mathbf{x},\mathbf{y},\omega) = \sum_{E_{ns}<0} \frac{\psi_{ns}(\mathbf{x})\bar{\psi}_{ns}(\mathbf{y})}{i\omega - E_{n,s}} + \text{contribution of the continuous spectrum.} \tag{22.19}$$

The perturbation scheme for calculating Z/Z_0 can be constructed by taking the quadratic form of a new field $\varphi(\mathbf{x},\tau)$ as in (22.18) as a free action, and considering the cubic, quartic and higher terms as perturbative. We can write the quadratic form as

$$Q[\varphi] = -\tfrac{1}{2}\sum_{\omega}\int \mathrm{d}\mathbf{x}\,\mathrm{d}\mathbf{y}\,\varphi(\mathbf{y},-\omega)A(\mathbf{x},\mathbf{y},\omega)\varphi(\mathbf{x},\omega), \qquad (22.20)$$

where

$$A(\mathbf{x},\mathbf{y},\omega) = -\nabla^2\delta(\mathbf{x}-\mathbf{y}) - 4\pi e^2\Pi(\mathbf{x},\mathbf{y},\omega). \qquad (22.21)$$

Here Π is the polarization operator. It is defined by the formula

$$\Pi(\mathbf{x},\mathbf{y},\omega) = \beta^{-1}\sum_{\omega_1,s} G_s(\mathbf{x},\mathbf{y},\omega_1)G_s(\mathbf{y},\mathbf{x},\omega+\omega_1). \qquad (22.22)$$

So we come to the following formula for $\ln(Z/Z_0)$:

$$\ln(Z/Z_0) = (\beta/8\pi)\int \mathrm{d}^3x(\nabla\phi)^2 - \tfrac{1}{2}\ln\det(A/A_0) + \sum_i D_i, \qquad (22.23)$$

where $\sum_i D_i$ is the sum of all vacuum diagrams, A is defined by (22.21) and (22.22), $A_0 = A|_{\Pi=0}$. If $T\to 0$ $(\beta\to\infty)$ we obtain the following formula for the ground state of a heavy atom:

$$E(Z) = \sum_{E_{ns}<0} E_{ns} - (8\pi)^{-1}\int \mathrm{d}^3x(\nabla\phi)^2$$
$$+ \lim_{\beta\to\infty}\frac{1}{2\beta}\ln\det(A/A_0) - \lim_{\beta\to\infty}\beta^{-1}\sum_i D_i. \qquad (22.24)$$

Here the E_{ns} are the negative eigenvalues of the Schrödinger equation (22.15), $\phi(\mathbf{x})$ is the electric potential field satisfying the Poisson equation (22.14).

The first two terms in (22.24) are nothing but the ground state energy in the Hartree approximation. If $\psi_{ns}(\mathbf{x})$ are the normalized eigenfunctions of the Schrödinger equation (22.15), corresponding to negative eigenvalues E_{ns}, we have

$$E_{ns} = E_{ns}\int \mathrm{d}^3x\bar{\psi}_{ns}(\mathbf{x})\psi_{ns}(\mathbf{x})$$
$$= \int \mathrm{d}^3x[(2m)^{-1}\nabla\bar{\psi}_{ns}(\mathbf{x})\nabla\psi_{ns}(\mathbf{x})$$
$$+(-\lambda + e\phi(\mathbf{x}) - Ze^2r^{-1})\bar{\psi}_{ns}(\mathbf{x})\psi_{ns}(\mathbf{x})]. \qquad (22.25)$$

Using the expression $\rho(\mathbf{x})$ via $\psi_{ns}(\mathbf{x})$:

$$\rho(\mathbf{x}) = \sum_{E_{ns}<0} \bar{\psi}_{ns}(\mathbf{x})\psi_{ns}(\mathbf{x}) \tag{22.26}$$

one can solve the Poisson equation (22.14) to find

$$\phi(\mathbf{x}) = e\int \frac{d^3y}{|\mathbf{x}-\mathbf{y}|} \sum_{E_{ns}<0} \bar{\psi}_{ns}(\mathbf{y})\psi_{ns}(\mathbf{y}). \tag{22.27}$$

So we obtain

$$\sum_{E_{ns}<0} E_{ns} - (8\pi)^{-1}\int d^3x(\nabla\phi)^2$$

$$= \sum_{E_{ns}<0}\int d^3x((2m)^{-1}\nabla\bar{\psi}_{ns}(\mathbf{x})\nabla\psi_{ns}(\mathbf{x}) - (\lambda + Ze^2r^{-1})\bar{\psi}_{ns}(\mathbf{x})\psi_{ns}(\mathbf{x})\psi_{ns}(\mathbf{x}))$$

$$+\frac{e^2}{2}\int \frac{d^3x d^3y}{|\mathbf{x}-\mathbf{y}|} \sum_{\substack{E_{ns}<0 \\ E_{n's'}<0}} \psi_{ns}(\mathbf{x})\bar{\psi}_{n's'}(\mathbf{y})\psi_{n's'}(\mathbf{y})\psi_{ns}(\mathbf{x}). \tag{22.28}$$

This is nothing but the well-known Hartree functional for the ground state energy of a heavy atom.

Other terms in (22.24) give us corrections to the Hartree approximation. For example, the Hartree–Fock exchange term is contained in $(2\beta)^{-1}\ln\det(A/A_0))$:

$$= (2\beta)^{-1}\ln(\det A/A_0)$$

$$= (2\beta)^{-1}\ln\det(1 + \nabla^{-2}4\pi e^2\Pi) \approx (2\beta)^{-1}\operatorname{Tr}(\nabla^{-2}\pi e^2\Pi)$$

$$= \lim_{\varepsilon\to+0}(2\beta^2)^{-1}\int \frac{e^2 d^3x d^3y}{|\mathbf{x}-\mathbf{y}|}\sum_s\left(\sum_\omega e^{i\omega\varepsilon}G_s(\mathbf{x},\mathbf{y},\omega)\right)\left(\sum_\omega e^{i\omega\varepsilon}G_s(\mathbf{y},\mathbf{x},\omega)\right). \tag{22.29}$$

We have

$$\lim_{\varepsilon\to+\infty}\beta^{-1}\sum_\omega e^{i\omega\varepsilon}G_s(\mathbf{x},\mathbf{y},\omega) = \sum_{E_{ns}<0}\frac{\psi_{ns}(\mathbf{x})\bar{\psi}_{ns}(\mathbf{y})}{e^{\beta E_{ns}}+1} + \int d\alpha\frac{\psi_{\alpha s}(\mathbf{x})\psi_{\alpha s}(\mathbf{y})}{e^{\beta E_\alpha}+1}, \tag{22.30}$$

where α corresponds to the contribution of the continuous spectrum. This contribution vanishes as $T\to 0$ ($\beta\to\infty$), and we obtain

$$\lim_{\beta\to\infty}\lim_{\varepsilon\to+0}\beta^{-1}\sum_\omega e^{i\omega\varepsilon}G_s(\mathbf{x},\mathbf{y},\omega) = \sum_{E_{ns}<0}\psi_{ns}(\mathbf{x})\bar{\psi}_{ns}(\mathbf{y}). \tag{22.31}$$

Substituting (22.31) into (22.29) we have

$$-\frac{e^2}{2}\int\frac{d^3x d^3y}{|\mathbf{x}-\mathbf{y}|}\sum_{E_{ns},E_{ms}0}\bar{\psi}_{ns}(\mathbf{x})\bar{\psi}_{ms}(\mathbf{y})\psi_{ns}(\mathbf{y})\psi_{ms}(\mathbf{x}). \tag{22.32}$$

This is the Hartree–Fock exchange correction to the Hartree functional (22.28).

The Poisson equation (22.14) turns into the well-known Thomas–Fermi equation for heavy atoms ($Z \gg 1$). We will prove this statement for the case of atoms with closed electron shells for such atoms ϕ dependent on r. The Schrödinger equation (22.15) may be replaced by a radial equation,

$$f'' + 2m(E - V_l)f = 0, \tag{22.33}$$

where f is defined by the formula

$$\psi = Y_{lm}(\theta, \varphi)r^{-1}f(r). \tag{22.34}$$

ψ is a solution of (22.15), $Y_{lm}(\theta, \varphi)$ is a spherical function satisfying the normalization condition

$$\int d\Omega |Y_{lm}(\theta, \varphi)|^2 = 4\pi. \tag{22.35}$$

V_l in (22.33) is the potential energy:

$$V_l(r) = V(r) + \frac{l(l+1)}{2mr^2} = -\lambda + e\phi(r) - \frac{Ze^2}{r} + \frac{l(l+1)}{2mr^2}. \tag{22.36}$$

It is easy to find a 'quasiclassical' solution to (22.33) of the form

$$f(r) = \exp i\int_{r_0}^{r} g(r)dr. \tag{22.37}$$

The equation for g is

$$ig' - g^2 + 2m(E - V_l) = 0. \tag{22.38}$$

We have

$$g = [2m(E - V_l) + ig']^{1/2}, \tag{22.39}$$

$$f(r) = \exp i\int_{r_0}^{r} dr[2m(E - V_l) + ig']^{1/2}. \tag{22.40}$$

Let us suppose $V_l(r)$ to be an analytic function in the vicinity of the real positive half-axes $r > 0$. In this case we have

$$\oint_C dr[2m(E - V_l) + ig']^{1/2} = 2\pi n. \tag{22.41}$$

Here C is a contour which encircles the 'classically allowed interval' $[r_1, r_2]$ $(V_l - E \geqslant 0$ if $r \in [r_1, r_2])$. See Fig. 22.1.

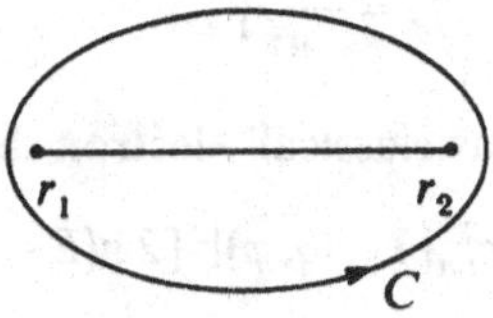

Fig. 22.1

The non-negative integer n is the number of zeros of $f(r)$. The quasiclassical approximation can be obtained from (22.41), if we suppose that $|g'| \ll |2m(V_l - E)| (r \in C)$. So we can write

$$g \approx [2m(E - V_l)]^{1/2}, \quad g' \approx -\frac{mV_l'}{[2m(E - V_l)]^{1/2}}, \tag{22.42}$$

$$[2m(E - V_l) + ig']^{1/2} \approx [2m(E - V_l)]^{1/2} + \frac{i}{4}\frac{V_l'}{V_l - E} \tag{22.43}$$

$$\frac{i}{4}\oint \frac{V_l' dr}{V_l - E} = \frac{i}{4} 2\pi i \cdot 2 = -\pi \tag{22.44}$$

and we come to the quasiclassical quantization condition

$$\int_{r_1}^{r_2} dr[2m(E - V_l)]^{1/2} = \pi(n + \tfrac{1}{2}), \tag{22.45}$$

which allows us to find the energy levels $E = E_{l,n}$. The quasiclassical wave-function is equal to

$$\psi_{l,m,n}(\mathbf{x}) = c_{l,n} Y_{l,m}(\theta, \varphi) r^{-1} [2m(E - V_l)]^{-1/4} \exp i \int_{r_1}^{r} [2m(E - V_l)]^{1/2} dr. \tag{22.46}$$

Here $c_{l,n}$ is a normalization constant which can be found from the normalization condition

$$1 = \int |\psi|^2 d^3x = 4\pi c_{l,n}^2 \int_{r_1}^{r_2} \frac{dr}{[2m(E - V_l)]^{1/2}}. \tag{22.47}$$

We can also express $c_{l,n}^2$ through $\Delta E = E_{l,n+1} - E_{l,n}$. From (22.45) we have

$$m\Delta E \int_{r_1}^{r_2} \frac{dr}{[2m(E - V_l)]^{1/2}} = \pi. \tag{22.48}$$

Comparing (22.47) with (22.48), we obtain

$$c_{l,n}^2 = \frac{m\Delta E_{l,n}}{4\pi^2}. \tag{22.49}$$

The expression for the 'quasiclassical' electron density

$$\rho(\mathbf{x}) = r^{-2} \sum_{l,m,n,s} c_{l,n}^2 |Y_{l,m}(\theta,\varphi)|^2 [2m(E - V_l)]^{-1/2} \tag{22.50}$$

may be rewritten as

$$\rho(\mathbf{x}) = (2\pi^2 r^2)^{-1} \sum_l (2l+1) \sum_n \frac{m\Delta E_{l,n}}{[2m(E - V_l)]^{1/2}} \tag{22.51}$$

if we use the identity

$$\sum_{-l \leqslant m \leqslant l} |Y_{l,m}(\theta,\varphi)|^2 = 2l+1 \tag{22.52}$$

and (22.49). We can approximately calculate $\sum_n$ by replacing it by the integral

$$\sum_n \frac{m\Delta E}{[2m(E - V_l)]^{1/2}} \approx \int_{V_l}^{0} \frac{m\mathrm{d}E}{[2m(E - V_l)]^{1/2}} = (-2mV_l)^{1/2}. \tag{22.53}$$

The sum $\sum_l$ is computed by similar means:

$$r^{-2} \sum_l (2l+1)(-2mV_l)^{1/2} \approx \int \mathrm{d}\frac{l(l+1)}{r^2}\left(-2mV(r) - \frac{l(l+1)}{r^2}\right)^{1/2}$$
$$= \tfrac{2}{3}(-2mV(r))^{3/2} = \tfrac{2}{3}[2m(\lambda + Ze^2 r^{-1} - e\phi(r))]^{3/2}. \tag{22.54}$$

We see that the Poisson equation (22.14) turns into the Thomas–Fermi equation

$$\nabla^2 \phi = -\frac{4e}{3\pi}[2m(\lambda + Zl^2 r^{-1} - e\phi(r))]^{3/2}. \tag{22.55}$$

We can write (22.55) in a simple form

$$x^{1/2}\frac{\mathrm{d}^2 U}{\mathrm{d}x^2} = -(1 - U(x))^{3/2} \tag{22.56}$$

by introducing a new function

$$e\phi = \frac{Ze^2}{r} U + \lambda \tag{22.57}$$

and a new dimensionless variable

$$r = ax, \quad a = \tfrac{1}{8}(6\pi)^{2/3} a_0 Z^{-1/3}, \quad a_0 = (ml^2)^{-1}, \tag{22.58}$$

where a_0 is the Bohr radius. We have to take the solution of (22.56) which satisfies the boundary conditions

$$U(0) = 0, \quad U(\infty) = 1. \tag{22.59}$$

For the theory of the Thomas–Fermi equation the reader is referred to Gombas (1950). Here we only show how to express the ground state energy in the Hartree approximation

$$\sum_{n,s} E_{ns} - (8\pi)^{-1} \int d^3x (\nabla\phi)^2 \tag{22.60}$$

in terms of $U(x)$. We have

$$(8\pi)^{-1} \int d^3x (\nabla\phi)^2 = \frac{Ze^2}{2} \int_0^\infty U_r^2 \, dr = 4(6\pi)^{-2/3} Z^{7/3} e^2 a_0^{-1} \int_0^\infty U_x^2 \, dx. \tag{22.61}$$

The sum $\sum_{n,s}$ may be evaluated approximately by the same method as when calculating $\rho(x)$

$$\sum_{n,s} E_{n,s} = 2\sum_{l,n}(2l+1)E_{l,n} \approx 2\sum_l (2l+1) \sum_n \frac{mE\Delta E}{\pi} \int \frac{dr}{[2m(E - V_l)]^{1/2}}, \tag{22.62}$$

$$\sum \frac{mE\Delta E}{[2m(E - V_l)]^{1/2}} \approx \int_{V_l}^0 \frac{mE dE}{[2m(E - V_l)]^{1/2}} = -\frac{1}{3m}(-2mV_l)^{3/2}, \tag{22.63}$$

$$\begin{aligned}\sum_l (2l+1)(-2mV_l)^{3/2} &\approx r^2 \int d\frac{l(l+1)}{r^2}\left(-2mV(r) - \frac{l(l+1)}{r^2}\right)^{3/2} \\ &= \tfrac{2}{5} r^2 (-2mV(r))^{5/2}.\end{aligned} \tag{22.64}$$

So we come to the expression:

$$\begin{aligned}\sum_{n,s} E_{ns} &= -\frac{4}{15\pi m} \int_0^\infty (-2mV(r))^{5/2} r^2 \, dr \\ &= -\frac{1}{15\pi^2 m} \int d^3x (-2mV(r))^{5/2}\end{aligned} \tag{22.65}$$

or

$$\sum_{n,s} E_{ns} = -\frac{16}{5} \frac{Z^{7/3} l^2 a_0^{-1}}{(6\pi)^{2/3}} \int_0^\infty dx \, x^{-1/2} (1 - U(x))^{5/2}. \tag{22.66}$$

As a result we find

$$E(Z) = -\frac{4Z^{7/3}e^2a_0^{-1}}{(6\pi)^{2/3}}\int_0^\infty dx\left(U_x^2 + \frac{4}{5}\frac{(1-U(x))^{5/2}}{x^{1/2}}\right). \tag{22.67}$$

We see that $E(Z)$ is proportional to $Z^{7/3}$. We also notice that the radial Thomas–Fermi equation (22.56) may be derived from the extremum condition for (22.67).

Corrections to (22.67) at large Z are small compared to this expression. For example, the Hartree–Fock exchange term (22.32) is of the order $Z^{5/3}e^2a_0^{-1}$ and the ratio of this term to the leading one (22.63) is of the order $Z^{-2/3} \ll 1$.

The next problem is that of evaluation of the 'collective level' energy corresponding to plasma oscillations. Such a level may be obtained as a pole of Green's function $D(1,2) = \langle \varphi(1)\varphi(2) \rangle$ of the electric potential field. The problem can be reduced to the solution of the Bethe–Solpeter type equation

$$D^{-1}(1,2)\psi = 0$$

or

$$-\nabla^2\psi(\mathbf{x},\omega) - 4\pi e^2\int d^3y\Pi(\mathbf{x},\mathbf{y},\omega)\psi(\mathbf{y},\omega) = 0. \tag{22.68}$$

Using a quasiuniform approximation, we replace Green's functions in

$$\Pi(\mathbf{x},\mathbf{y},\omega) = \sum_{\omega_1,s} G_s(\mathbf{x},\mathbf{y},\omega_1)G_s(\mathbf{y},\mathbf{x},\omega+\omega_1) \tag{22.69}$$

by Green's functions of the uniform electron gas with density $\rho = (3\pi^2)^{-1} k_F^3$, which is characteristic of the central part of a heavy atom. Collective excitation in this approximation are nothing but plasma oscillations, their frequency being

$$\omega_0 = \left(\frac{4\pi e^2\rho}{m}\right)^{1/2}. \tag{22.70}$$

It is clear that this formula gives us only the order of magnitude of the plasma frequency in a heavy atom. A more exact approach to the problem of collective levels and their damping must be based on (22.68).

23

Functional integral approach to the theory of model Hamiltonians

In this section we outline the functional integral approach to the mathematically rigorous theory of the so-called model Hamiltonians system such as the Bardeen–Cooper–Schriffer (BCS) model in superconductivity (Bardeen, Cooper & Schriffer, 1957) or the Dicke model of superradiation. The problem of developing a rigorous theory of such models was put forward by Bogoliubov, Zubarev & Tserkovnikov (1960). This problem was solved for the BCS model by Bogoliubov (1960), who developed the so-called approximation Hamiltonian method. This method may be applied to the Dicke model as well. A rigorous theory of the Dicke model was suggested by Hepp & Lieb (1973). They have shown that there exists an exact solution of the model in the thermodynamic limit. The superradiation phase transition in this model was also described by these authors.

The approximation Hamiltonian method was applied to the Dicke model by Bogoliubov Jr (1974). Rigorous results for free energy and boson averages were obtained (see Bogoliubov Jr *et al.*, 1981).

Functional integral methods were also applied to the Dicke model. Moshchinsky & Fedianin (1977) obtained asymptotics of Z/Z_0, where Z is the partition function of the model and Z_0 is the partition function of the corresponding free system. Kirianov and Yarunin (1980) investigated the Bose excitation spectrum of the system below the phase transition point in the superradiation state.

In this section we will prove the asymptotics of Z/Z_0 for the Dicke model with a single mode of the radiation field (Popov & Fedotov, 1982). This proof can be carried over to an arbitrary finite number of Bose modes. At the end of the section we discuss the functional integral approach to the BCS model (Popov, 1980). Here we also get the formula for Z/Z_0 without its rigorous justification.

The Dicke model has the following Hamiltonian:

$$H = \omega_0 b^+ b + \sum_{i=1}^{N} \frac{\Omega}{2}(\psi_{2i}^+\psi_{2i} - \psi_{1i}^+\psi_{1i})$$
$$+ gN^{-1/2} \sum_{i=1}^{N} (b^+\psi_{1i}^+\psi_{2i} + b\psi_{2i}^+\psi_{1i}). \tag{23.1}$$

This model describes the one-mode Bose field b, b^+ interacting with the Fermi fields of atoms ψ_{ai}, ψ_{ai}^+ ($i = 1, 2, \ldots, N$ being the number of atoms); the index $a = 1, 2$ shows whether the atom is in the ground state ($a = 1$) or in an excited state ($a = 2$).

Let us write down the quotient Z/Z_0 as a formal quotient of two functional integrals

$$\frac{Z}{Z_0} = \frac{\int e^S D\mu}{\int e^{S_0} D\mu}. \tag{23.2}$$

Here S is a Euclidean action corresponding to the Hamiltonian (23.1):

$$S = \int_0^\beta d\tau \left(b^*(\tau)\partial_\tau b(\tau) + \sum \psi_{ai}^*(\tau)\partial_\tau \psi_{ai}(\tau) - H(b, b^*(\psi_{ai}, \psi_{ai}^*)) \right), \tag{23.3}$$

S_0 is the free system action, $D\mu$ is the measure of integration. The functional integrals in (23.2) are the integrals with respect to the complex functions $b(\tau), b^*(\tau)$, and Grassmann Fermi fields $\psi_{ai}(\tau), \psi_{ai}^*(\tau)$, subject to the following (anti) periodic conditions $b(\beta) = b(0), \psi_{ai}(\beta) = -\psi_{ai}(0) (\tau \in [0, \beta])$. Expanding the fields $b(\tau)$, $\psi_{ai}(\tau)$ into Fourier series

$$b(\tau) = \beta^{-1/2} \sum_\omega b(\omega) e^{i\omega\tau}, \quad \psi_{ai}(\tau) = \sum_p \psi_{ai}(p) e^{ip\tau}, \tag{23.4}$$

where $\omega = 2\pi n/\beta$, $p = (2n+1)\pi/\beta$, we may write the functional integral measure $D\mu$ in the form

$$D\mu = \prod_\omega db^*(\omega) db(\omega) \prod_{p,i,a} d\psi_{ai}^*(p) d\psi_{ai}. \tag{23.5}$$

The quotient of two integrals in (23.2) can be understood to be the limit of the quotient of finite dimensional integrals, by making the cutoff $|\omega| < \omega_B, |p| < \omega_F$ and then considering the limit $\omega_B, \omega_F \to \infty$.

The integrals with respect to Fermi fields are Gaussian, and we may integrate over these variables:

$$\frac{Z}{Z_0} = \frac{\int e^{S_0(b)} \det^N M(b, b^*) D\mu(b)}{\int e^{S_0(b)} \det M(0, 0) D\mu(b)}. \tag{23.6}$$

Here

$$S_0(b) = \sum_\omega (i\omega - \omega_0) b^*(\omega) b(\omega), \tag{23.7}$$

M is an operator with the following elements:

$$M_{pq} = \begin{pmatrix} (ip + \Omega/2)\delta_{pq}, & g(\beta N)^{-1/2} b^*(q-p) \\ g(\beta N)^{-1/2} b(p-q), & (ip - \Omega/2)\delta_{pq} \end{pmatrix}. \qquad (23.8)$$

Before the limit ω_B, $\omega_F \to \infty$ is taken, all integrals in (23.6) are finite dimensional and convergent, due to the exponential decay of $\exp S_0(b)$ as $|b(\omega)|^2 \to \infty$ and the fact that $\det^N M(b, b^*)$ is a polynomial function of $b(\omega), b^*(\omega)$. We can transfer the factor $\det^N M(0,0)$, which does not depend on b, b^*, from the denominator to the numerator of (23.6).

Let us make a change of variables

$$b(\omega) \to [\pi/(\omega_0 - i\omega)]^{1/2} b(\omega), \quad b^*(\omega) \to [\pi/(\omega_0 - i\omega)]^{1/2} b^*(\omega). \qquad (23.9)$$

The right-hand side in (23.9) are not conjugate to each other. Nevertheless, it is not difficult to justify (23.9) if we introduce polar coordinates instead of $b(\omega), b^*(\omega)$: $b(\omega) = (\rho(\omega))^{1/2} e^{i\varphi(\omega)}$, $b^*(\omega) \to (\rho(\omega))^{1/2} e^{-i\varphi(\omega)}$ and then perform a complex rotation of the integration counter when integrating with respect to $\rho(\omega)$: $\rho(\omega) \to \rho(\omega)\ [\pi/(\omega_0 - i\omega)]^{1/2}$. After this transformation we may come back to $b(\omega)$, $b^*(\omega)$. As a result we get the formula:

$$\frac{Z}{Z_0} = \int D\mu(b) \exp\left(-\pi \sum_{\omega} b^*(\omega) b(\omega) \right) \det^N (I + A). \qquad (23.10)$$

Here

$$\det(I + A) = \det M^{-1/2}(0,0) M(b, b^*) M^{-1/2}(0,0) \qquad (23.11)$$

and the operator A is defined as follows:

$$A_{pq} = \begin{pmatrix} 0 & B_{pq} \\ -C_{pq} & 0 \end{pmatrix}, \qquad (23.12)$$

where

$$B_{pq} = g\left(\frac{\pi}{\beta N}\right)^{1/2} b^*(q-p) \left[\left(\frac{\Omega}{2} + ip\right)(\omega_0 - i(q-p))\left(\frac{\Omega}{2} - iq\right) \right]^{-1/2}, \qquad (23.13)$$

$$C_{pq} = g\left(\frac{\pi}{\beta N}\right)^{1/2} b(p-q) \left[\left(\frac{\Omega}{2} - ip\right)(\omega_0 - i(p-q))\left(\frac{\Omega}{2} + iq\right) \right]^{-1/2}. \qquad (23.14)$$

The integral in the denominator of (23.6) turns out to be equal to unity:

$$\int D\mu(b) \exp\left(-\pi \sum_{\omega} b^*(\omega) b(\omega) \right) = 1, \qquad (23.15)$$

In (23.10) we may go to the limit $\omega_B, \omega_F \to \infty$ and then instead of a formal quotient of two (infinite) functional integrals we shall have only one finite functional integral. This representation turns out to be very useful for obtaining the asymptotic formula for Z/Z_0 at large N. These asymptotes are defined by the formulae of the stationary phase method. There exists only one stationary phase point at $T > T_c$. If $T < T_c$, we have a circle of a stationary phase $|b(0)|^2 = \rho_0, b(\omega) = b^*(\omega) = 0$ if $\omega \neq 0$. There also exists an interpolation formula between these asymptotes. Below we shall obtain all these asymptotes and show how one can justify them rigorously.

We begin with the investigation of the integral (23.10) for $T > T_c$. First of all let us show that this integral converges. We use the estimate:

$$|\det(I + A)| \leqslant \exp(\operatorname{Re}\operatorname{tr} A + \tfrac{1}{2}\operatorname{tr} AA^+). \tag{23.16}$$

The operator A has the form (23.12), and we find

$$\operatorname{tr} A = 0, \quad \operatorname{tr} AA^+ = \operatorname{tr} BB^+ + \operatorname{tr} CC^+. \tag{23.17}$$

We obtain the estimate:

$$\frac{Z}{Z_0} \leqslant I_0 = \int \mathrm{D}\mu(b) \exp\left(-\pi \sum_{\omega} (1 - a_0(\omega)) b^*(\omega) b(\omega)\right)$$

$$= \prod_{\omega} (1 - a_0(\omega))^{-1}. \tag{23.18}$$

Here

$$a_0(\omega) = g^2 \beta^{-1} (\omega^2 + \omega_0^2)^{-1/2} \sum_{q-p=\omega} \left[\left(p^2 + \frac{\Omega^2}{4}\right)\left(q^2 + \frac{\Omega^2}{4}\right)\right]^{-1/2}. \tag{23.19}$$

We have $0 < a_0(\omega) \leqslant a_0(0)$, and $a_0(\omega) = O(\omega^{-2} \ln \omega)$. This implies the existence of the integral in (23.18) and the convergence of the infinite product on the right-hand side of (23.18), if $a_0(0) < 1$. The condition $a_0(0) = 1$ is nothing but the well-known equation for the transition temperature in the Dicke model:

$$a_0(0) = \frac{g^2}{\omega_0 \Omega} \tanh \frac{\Omega}{4T_c} = 1. \tag{23.20}$$

So, if $T > T_c$, we have $a_0(0) < 1$, the integral (23.10) exists and does not exceed I_0.

Now let us show that, at $T > T_c$, this integral may be evaluated by the stationary phase method if we make the replacement

$$\det^N(I+A)=\det^N(I+BC)\approx \exp N \operatorname{tr} BC. \tag{23.21}$$

In order to justify this substitute and to estimate the error, we divide all the functional space into two domains C_1 and C_2 according to whether

$$\operatorname{tr} BC(BC)^+ \leqslant (4N)^{-1} \quad C_1$$

or

$$\operatorname{tr} BC(BC)^+ \geqslant (4N)^{-1} \quad C_2. \tag{23.22}$$

We denote

$$K_N=\det^N(I+A)-\exp(N\operatorname{tr} BC). \tag{23.23}$$

Consider the following identity:

$$\begin{aligned}\frac{Z}{Z_0}=&\int \mathrm{D}\mu(b)\exp\left(-\pi\sum_\omega |b(\omega)|^2+N\operatorname{tr} BC\right)\\&+\int_{C_1}\mathrm{D}\mu(b)K_N\exp\left(-\pi\sum_\omega|b(\omega)|^2\right)\\&+\int_{C_2}\mathrm{D}\mu(b)[\det^N(I+A)\\&-\exp N\operatorname{tr} BC]\exp\left(-\pi\sum_\omega|b(\omega)|^2\right).\end{aligned} \tag{23.24}$$

The first term on the right-hand side of (23.23) is equal to

$$\int \mathrm{D}\mu(b)\exp\left(-\pi\sum_\omega|b(\omega)|^2(1-a(\omega))\right)=\prod_\omega(1-a(\omega))^{-1}, \tag{23.25}$$

which is the main term in the asymptotes of Z/Z_0. Here

$$a(\omega)=g^2(\omega_0-\mathrm{i}\omega)^{-1}\sum_{q-p=\omega}\left(\frac{\Omega}{2}+\mathrm{i}p\right)^{-1}\left(\frac{\Omega}{2}-\mathrm{i}q\right)^{-1}=\frac{g^2\tanh\dfrac{\beta\Omega}{4}}{(\omega_0-\mathrm{i}\omega)(\Omega-\mathrm{i}\omega)} \tag{23.26}$$

we have $|a(\omega)|\leqslant a_0(\omega)\leqslant a_0(0)=a(0)$. This implies that the integral in (23.25) does really exist if $T>T_c$, and the infinite product in the right-hand side of (23.25) converges for $T>T_c$.

The third term on the right-hand side (23.24) does not exceed

$$2\int_{C_2}\mathrm{D}\mu(b)\exp\left(-\pi\sum_\omega|b(\omega)|^2(1-a_0(\omega))\right) \tag{23.27}$$

since both $|\det(I + A)|^N$ and $\exp N \operatorname{tr} BC$ do not exceed

$$\exp\frac{N}{2}\operatorname{tr}(BB^+ + CC^+) = \exp\left(\pi\sum_\omega |b(\omega)|^2 a_0(\omega)\right). \tag{23.28}$$

Now we can estimate (23.27) via

$$8N\int_{C_2} \mathrm{D}\mu(b)\operatorname{tr}(BC(BC)^+)\exp\left(-\pi\sum_\omega |b(\omega)|^2(1 - a_0(\omega))\right) \tag{23.29}$$

using the second condition (23.22) defining C_2.

Now we estimate the second term in the right-hand side of (23.24). We have

$$K_N = \prod_i (1 + \lambda_i)^N - \prod_i e^{N\lambda_i} = (e^X - 1)\prod_i e^{N\lambda_i}, \tag{23.30}$$

where λ_i are the eigenvalues of the operator BC,

$$X = N\sum_i (\ln(1 + \lambda_i) - \lambda_i). \tag{23.31}$$

In the domain C_1 we have

$$\sum_i |\lambda_i|^2 \leqslant \operatorname{tr} BC(BC)^+ \leqslant (4N)^{-1} \leqslant 1/4. \tag{23.32}$$

So all the $|\lambda_i|$ do not exceed 1/2, and we may estimate X as follows:

$$|X| \leqslant N\sum_i |\lambda_i|^2 \leqslant N \operatorname{tr} BC(BC)^+ \leqslant 1/4. \tag{23.33}$$

Then we have the following estimates:

$$\left|\prod_i e^{N\lambda_i}\right| \leqslant \exp\left(\pi\sum_\omega |b(\omega)|^2 a_0(\omega)\right), \tag{23.34}$$

$$|e^X - 1| \leqslant e^{|X|} - 1 \leqslant |X| e^{|X|} \leqslant e^{1/4} N \operatorname{tr} BC(BC)^+. \tag{23.35}$$

These estimates allow us to estimate the second term plus the third term in (23.24) via

$$\begin{aligned} &8N\int_{C_2} \operatorname{tr}(BC(BC)^+)\exp\left(-\pi\sum_\omega |b(\omega)|^2(1 - a_0(\omega))\right)\mathrm{D}\mu(b) \\ &+ e^{1/4}N\int_{C_2} \operatorname{tr}(BC(BC)^+)\exp\left(-\pi\sum_\omega |b(\omega)|^2(1 - a_0(\omega))\right)\mathrm{D}\mu(b) \\ &\leqslant 8N\int \operatorname{tr}(BC(BC)^+)\exp\left(-\pi\sum_\omega |b(\omega)|^2(1 - a_0(\omega))\right)\mathrm{D}\mu(b). \end{aligned} \tag{23.36}$$

Now using the inequality (where ε is an arbitrary positive number)

$$8N \operatorname{tr} BC(BC)^+ \leqslant 2N \operatorname{tr}(B^+B + C^+C)^2$$
$$\leqslant \frac{16}{\varepsilon^2 N} \exp \tfrac{1}{2}\varepsilon N \operatorname{tr}(B^+B + C^+C)$$
$$= \frac{16}{\varepsilon^2 N} \exp\left(\varepsilon \sum_\omega |b(\omega)|^2 a_0(\omega)\right) \tag{23.37}$$

we may estimate the corrective term to (23.18) by

$$\frac{16}{\varepsilon^2 N} \prod_\omega (1 - (1+\varepsilon)a_0(\omega))^{-1}. \tag{23.38}$$

We see that if T is separated from T_c ($T \geqslant T_c + \delta, \delta > 0$) then

$$\frac{Z}{Z_0} = \prod_\omega (1 - a(\omega))^{-1} + O(N^{-1}). \tag{23.39}$$

Now let us proceed to the case $T < T_c$. Here we have the same formula for Z/Z_0 (23.10) as we had for $T > T_c$. The proof of convergence of this integral must be modified as well as the method for obtaining the asymptotes. The point is that for $T < T_c$, the main contribution into the integral comes from the vicinity of some stationary phase circle $b(\omega) = b^*(\omega) = 0$ $(\omega \neq 0)$, $|b(0)|^2 = N\beta\omega_0\pi^{-1}\Delta^2$, where Δ is the condensate density. Let us go to polar coordinates in the integral with respect to $b(0)$, $b^*(0)$ ($b(0) = \rho e^{i\varphi}$, $b^*(0) = \rho e^{i\varphi}$) and then make the changes $b(\omega) \to b(\omega)e^{i\varphi}, b^*(\omega) \to b^*(\omega)e^{-i\varphi}$ in the integral over $b(\omega), b^*(\omega)$ with $\omega \neq 0$. (φ is the phase of $b(0)$). After this change of variables we may write the integral (23.10) in the following form:

$$\pi \iint \left[d\rho^2 \prod_{\omega \neq 0} db^*(\omega) db(\omega) \right] \det{}^N(I + A) \exp\left(-\pi \sum_\omega |b(\omega)|^2 \right), \tag{23.40}$$

where $b(0) = b^*(0) = \rho$ and the integrand does not depend on φ (the phase of the variable $b(0)$).

The integrand (23.40) is maximal at $b(\omega) = b^*(\omega) = 0$, $(\omega \neq 0)$, $\rho^2 = \rho_0^2 = N\beta\omega_0\pi^{-1}\Delta^2$, where Δ can be found from the extremum condition. The explicit form of this condition is

$$1 = \frac{g^2}{\omega_0[\Omega^2 + 4g^2\Delta^2]^{1/2}} \tanh \frac{\beta}{4} [\Omega^2 + 4g^2\Delta^2]^{1/2}. \tag{23.41}$$

We can single out the extremal value of the integrand and write down Z/Z_0 as follows:

$$\frac{Z}{Z_0} = \pi e^{aN} J \tag{23.42}$$

Here

$$a = 2\ln\left[\frac{\cosh(\beta\Omega_\Delta/4)}{\cosh(\beta\Omega/4)}\right] - \Delta^2\beta\omega_0, \tag{23.43}$$

where

$$\Omega_\Delta^2 = \Omega^2 + 4g^2\Delta^2. \tag{23.44}$$

$$J = \iint\left[\mathrm{d}\rho^2 \prod_{\omega\neq 0} \mathrm{d}b^*(\omega)\mathrm{d}b(\omega)\right]\det{}^N(I+B)\exp\left(-\pi\sum_\omega |b(\omega)|^2 + N\beta\omega_0^2\Delta^2\right), \tag{23.45}$$

$$\det(I+B) = \det M^{-1/2}(\rho_0,\rho_0)M(b,b^*)M^{-1/2}(\rho_0,\rho_0). \tag{23.46}$$

Here aN is equal to

$$aN = -\beta(F_\Delta - F_0), \tag{23.47}$$

where $F_\Delta - F_0$ is the difference between the free energy of the Dicke model in the superradiant state and the free energy of the system without interaction.

It turns out that one can give a rigorous proof of the existence of J (23.45), and can obtain its asymptotes, as $N\to\infty$. The asymptotic behaviour for $T < T_c$ is given by

$$Z/Z_0 = Ae^{aN}N^{1/2}(1 + O(N^{-1/2})). \tag{23.48}$$

Here a is defined by (23.43) and

$$A = g^{-1}(\pi\beta\omega_0)^{1/2}\Omega_\Delta\left(1 - \frac{\beta\Omega_\Delta}{2\sinh\beta\Omega_\Delta/2}\right)^{-1} \cdot \prod_{\omega>0} \det{}^{-1}\begin{pmatrix} 1-\alpha(\Delta,\omega), & \beta(\Delta,\omega) \\ \beta(\Delta,\omega), & 1-\alpha(\Delta,\omega)\end{pmatrix}. \tag{23.49}$$

where

$$\alpha(\Delta,\omega) = \frac{g^2}{\beta(\omega_0 - \mathrm{i}\omega)}\sum_{p-q=\omega}\frac{(\Omega/2 - \mathrm{i}p)(\Omega/2 + \mathrm{i}q)}{(\Omega^2/4 + p^2 + g^2\Delta^2)(\Omega^2/4 + q^2 + g^2\Delta^2)}$$
$$= \frac{g^2\tanh(\beta\Omega_\Delta/4)(\Omega^2 + 2g^2\Delta^2 + \Omega\mathrm{i}\omega)}{(\omega_0 - \mathrm{i}\omega)\Omega_\Delta(\Omega_\Delta^2 + \omega^2)}$$

$$= \frac{\omega_0(\Omega^2 + 2g^2\Delta^2 + \Omega i\omega)}{(\omega_0 - i\omega)\Omega_\Delta(\Omega_\Delta^2 + \omega^2)}$$

and

$$\beta(\Delta, \omega) = \frac{g^2}{\beta(\omega_0^2 + \omega^2)^{1/2}} \sum_{q-p=\omega} g^2\Delta^2\left(\frac{\Omega^2}{4} + p^2 + g^2\Delta^2\right)^{-1}$$

$$\cdot\left(\frac{\Omega^2}{4} + q^2 + g^2\Delta^2\right)^{-1}$$

$$= \frac{2g^4\Delta^2 \tanh(\beta\Omega_\Delta/4)}{(\omega_0^2 + \omega^2)^{1/2}\Omega_\Delta(\Omega_\Delta^2 + \omega^2)}$$

$$= \frac{2g^2\Delta^2\omega_0}{(\omega_0^2 + \omega^2)^{1/2}(\Omega_\Delta^2 + \omega^2)}. \tag{23.50}$$

The details of a rigorous justification of (23.48) can be found in the paper by Popov & Fedotov (1982). The idea is that the main contribution into the functional integral (23.45) comes from the neighbourhood of the point where $b(\omega) = b^*(\omega) = 0, \rho = \rho_0$. We can estimate the integral over the region of functional space outside the above-mentioned neighbourhood and this enables us to show that the integral over this neighbourhood is equal approximately to $Ae^{aN}N^{1/2}$.

It is also possible to derive a formula interpolating between (23.39) and (23.48) which is valid in the vicinity of the transition, namely in the region

$$\frac{N^{1/2}(T - T_c)}{T_c} \sim 1. \tag{23.51}$$

This interpolation formula is the following one:

$$\frac{Z}{Z_0} = Ae^{\tau^2/2}\int_{-\infty}^{\tau} e^{-t^2/2}\,dt, \tag{23.52}$$

where

$$\tau = \frac{(1 - a(0))\Omega(N\beta\omega_0)^{1/2}}{g(2 - \beta\Omega/\sinh(\beta\Omega/2))^{1/2}} \sim \frac{N^{1/2}(T - T_c)}{T_c}, \tag{23.53}$$

$$A = \frac{(\beta\omega_0)^{1/2}\Omega}{g(2 - \beta\Omega/\sinh(\beta\Omega/2))^{1/2}} \prod_{\substack{\omega \neq 0 \\ T = T_c}} (1 - a(\omega))^{-1}. \tag{23.54}$$

The formulae (23.39), (23.48) and (23.52) are obtained and rigorously justified in the framework of the functional integration method. They can be difficult to obtain through other approaches. Here we have outlined a rigorous proof only for (23.39), which is valid for $T > T_c + \delta$. The proof of (23.48) and (23.52) was given by Popov & Fedotov (1981).

The same approach can be applied to the BCS model (Popov, 1980). Its Hamiltonian is

$$H = \sum_{\mathbf{k},s} T(\mathbf{k}) a_s^+(\mathbf{k}) a_s(\mathbf{k}) - \frac{1}{V} \sum_{\mathbf{k},\mathbf{k}'} \lambda(\mathbf{k})\lambda(\mathbf{k}') a_+^+(\mathbf{k}) a_-^+(-\mathbf{k}) a_-(-\mathbf{k}') a_+(\mathbf{k}'). \quad (23.55)$$

Here $a_s^+(\mathbf{k}), a_s(\mathbf{k})$ are the Fermi creation and annihilation operators of particles with momentum $\mathbf{k}$ and spin $s = \pm$. For instance, we can take

$$T(\mathbf{k}) = \frac{k^2}{2m} - \mu = \frac{k^2 - k_F^2}{2m} \quad (23.56)$$

and assume $\lambda(\mathbf{k})$ to be constant in a narrow shell about the Fermi sphere $k^2 = k_F^2$ and to vanish outside this shell. We will suppose $T(\mathbf{k}), \lambda(\mathbf{k})$ to be even functions.

When using the functional integration method, we have to deal with the action functional

$$\begin{aligned} S = & \sum_{p,s} (i\omega - T(\mathbf{k})) a_s^*(p) a_s(p) \\ & + (\beta V)^{-1} \sum_{\substack{\mathbf{k}_1 + \mathbf{k}_2 = \mathbf{k}_3 + \mathbf{k}_4 = 0 \\ \omega_1 + \omega_2 = \omega_3 + \omega_4}} \lambda(\mathbf{k}_1)\lambda(\mathbf{k}_4) a_+^*(p_1) a_-^*(p_2) a_-(p_3) a_+(p_4), \end{aligned} \quad (23.57)$$

where $a_s(p) = a_s(\mathbf{k}, \omega), a_s^*(p) = a_s^*(\mathbf{k}, \omega)$ are generators of the Grassmann algebra, $\omega = (2n+1)\pi/\beta$ are Fermi frequencies.

We can express the quotient Z/Z_0 as a formal quotient of functional integrals

$$\frac{Z}{Z_0} = \frac{\int e^S \, Da^* \, Da}{\int e^{S_0} \, Da^* \, Da}. \quad (23.58)$$

Here S is equal to (23.57), S_0 is the action of a free system action, which can be obtained from (23.57) at $\lambda(\mathbf{k}) = 0$,

$$Da^* Da = \prod_{p,s} da_s^*(p) \, da_s(p). \quad (23.59)$$

Now we can insert the following functional integral over the auxiliary Bose field in both the numerator and the denominator of (23.58):

$$\int \exp\left(-\sum_{\omega} c^*(\omega) c(\omega) \right) \prod_{\omega} dc^*(\omega) dc(\omega). \quad (23.60)$$

Then we perform the shift transformations

$$c(\omega) \to c(\omega) - (\beta V)^{-1/2} \sum_{\substack{\omega_1+\omega_2=\omega \\ \mathbf{k}_1+\mathbf{k}_2=0}} \lambda(\mathbf{k}_1) a_-(p_2) a_+(p_1),$$

$$c^*(\omega) \to c^*(\omega) - (\beta V)^{-1/2} \sum_{\substack{\omega_1+\omega_2=\omega \\ \mathbf{k}_1+\mathbf{k}_2=0}} \bar{\lambda}(\mathbf{k}_1) a_+^*(p_1) a_-^*(p_2). \quad (23.61)$$

These transformations allow us to cancel the interation form of the action functional, and so we obtain the following action depending on both Fermi and Bose fields:

$$-\sum_{\omega} c^*(\omega)c(\omega) + \sum_{p,s} (\mathrm{i}\omega - T(\mathbf{k})) a_s^*(p) a_s(p)$$
$$+ (\beta V)^{-1/2} \sum_{\substack{\omega_1+\omega_2=\omega \\ \mathbf{k}_1+\mathbf{k}_2=0}} [\lambda(\mathbf{k}_1) a_-(p_2) a_+(p_1) c^*(\omega) + \bar{\lambda}(\mathbf{k}_1) a_+^*(p_1) a_-^*(p_2) c(\omega)]. \quad (23.62)$$

Now we may integrate with respect to Fermi fields in both the numerator and the denominator to obtain

$$\frac{Z}{Z_0} = \frac{\int \exp\left(-\sum_{\omega} c^*(\omega)c(\omega)\right) \det A[c(\omega), c^*(\omega)]/A[0,0] \prod_{\omega} \mathrm{d}c^*(\omega)\mathrm{d}c(\omega)}{\int \exp\left(-\sum_{\omega} c^*(\omega)c(\omega)\right) \prod_{\omega} \mathrm{d}c^*(\omega)\mathrm{d}c(\omega)} \quad (23.63)$$

where A is an operator defined by

$$A_{p_1 p_2} = \delta_{\mathbf{k}_1 \mathbf{k}_2} \begin{pmatrix} (\mathrm{i}\omega_1 - T(\mathbf{k}_1))\delta_{\omega_1,\omega_2}, & (\beta V)^{-1/2} \bar{\lambda}(\mathbf{k}_1) c(\omega_1 - \omega_2) \\ (\beta V)^{-1/2} \lambda(\mathbf{k}_1) c^*(\omega_2 - \omega_1), & (\mathrm{i}\omega_1 + T(\mathbf{k}_1))\delta_{\omega_1 \omega_2} \end{pmatrix}. \quad (23.64)$$

Let A_0 be the operator A evaluated at $c = c^* = 0$,

$$A = A_0 + u. \quad (23.65)$$

So we may write

$$\det A/A_0 = \exp \operatorname{Tr} \ln (A/A_0) = \exp \operatorname{Tr} \ln (I + Gu), \quad (23.66)$$

where $G = A_0^{-1}$.

Now it is easy to obtain the asymptotes of Z/Z_0 for $T > T_c$. It is natural to suppose that the main contribution into the functional integral comes from a neighbourhood of the zero point $c(\omega) = c^*(\omega) = 0$ of the functional space.

That is why we can obtain a formula for Z/Z_0, by retaining only the first term in the expansion:

$$\operatorname{Tr}\ln(I+Gu) = -\sum_{n=1}^{\infty}(2n)^{-1}\operatorname{Tr}(Gu)^{2n}. \tag{23.67}$$

Here we take into account that only even powers of Gu give a contribution into $\operatorname{Tr}\ln$. We have

$$-\tfrac{1}{2}\operatorname{Tr}(Gu)^2 = -\sum_{\omega} c^*(\omega)c(\omega)\frac{1}{\beta V}\sum_{\mathbf{k}_1,\omega_1}\frac{|\lambda(\mathbf{k}_1)|^2}{(i\omega_1 - T(\mathbf{k}))(i\omega_1 - i\omega - T(-\mathbf{k}))}$$

$$= \sum_{\omega} c^*(\omega)c(\omega)\frac{1}{V}\sum_{\mathbf{k}}\frac{|\lambda(\mathbf{k})|^2\tanh\frac{\beta}{2}T(\mathbf{k})}{2T(\mathbf{k}) - i\omega}. \tag{23.68}$$

Now if we substitute $\exp(-\frac{1}{2}\operatorname{Tr}(Gu)^2)$ for $\det A/A_0$ in the functional integral in (23.63), we come to the formula

$$\frac{Z}{Z_0} \approx \sum_{\omega}\left(1 - \frac{1}{V}\sum_{\mathbf{k}}\frac{|\lambda(\mathbf{k})|^2\tanh\frac{1}{2}\beta T(\mathbf{k})}{2T(\mathbf{k}) - i\omega}\right)^{-1}. \tag{23.69}$$

This infinite product is not absolutely convergent. It can be shown that (23.69) must be interpreted as the limit

$$\frac{Z}{Z_0} = \lim_{\varepsilon\to+0}\prod_{\omega}\left(1 - \frac{e^{i\omega\varepsilon}}{V}\sum_{\mathbf{k}}\frac{|\lambda(\mathbf{k})|^2\tanh\frac{1}{2}\beta T(\mathbf{k})}{2T(\mathbf{k}) - i\omega}\right)^{-1}. \tag{23.70}$$

This result makes sense unless the central multiplier in $\prod_{\omega}$ (corresponding to $\omega = 0$) becomes infinite. The condition for the central multiplier to be infinite is

$$1 = \frac{1}{2V}\sum_{k}\frac{|\lambda(\mathbf{k})|^2}{T(\mathbf{k})}\tanh\tfrac{1}{2}\beta T(\mathbf{k}). \tag{23.71}$$

This is just the equation for the phase transition temperature in the BCS model.

It is also possible to obtain Z/Z_0 below T_c. Here the main contribution into the integral with respect to $c(\omega), c^*(\omega)$ comes from a neighbourhood of the circle $|c(0)|^2 = \Delta^2\beta V$, where Δ^2 is defined by

$$1 = \frac{1}{2V}\sum_{k}\frac{|\lambda(\mathbf{k})|^2}{\varepsilon(\mathbf{k})}\tanh\tfrac{1}{2}\beta\varepsilon(\mathbf{k}),$$

$$\varepsilon^2(\mathbf{k}) = T^2(\mathbf{k}) + \Delta^2|\lambda(\mathbf{k})|^2. \tag{23.72}$$

It can be shown that

$$\frac{Z}{Z_0} \approx (\exp \beta V(p - p_0)) \left(\pi\beta V \Big/ \frac{1}{2\beta V} \sum_{\mathbf{k},\omega} \frac{|\lambda(\mathbf{k})|^4}{(\omega^2 + \varepsilon^2(\mathbf{k}))^2} \right)^{1/2}$$

$$\times \lim_{\varepsilon \to +0} \prod_{\omega > 0} \det{}^{-1} \begin{pmatrix} 1 - e^{i\omega\varepsilon}\alpha(\omega), & \beta(\omega) \\ \beta(\omega), & 1 - e^{-i\omega\varepsilon}\alpha(-\omega) \end{pmatrix}. \quad (23.73)$$

Here p is the pressure in the model system at $T < T_c$, and p_0 is the pressure of the corresponding free system. We have

$$p - p_0 = -\Delta^2 + \frac{2}{\beta V} \sum_{\mathbf{k}} \ln \frac{\cosh \frac{\beta}{2} [T^2(\mathbf{k}) + \Delta^2 |\lambda(\mathbf{k})|^2]^{1/2}}{\cosh \frac{\beta}{2} T(\mathbf{k})}. \quad (23.74)$$

Choosing Δ from the maximum condition for (23.74) we obtain (23.72). The functions $\alpha(\omega)$ and $\beta(\omega)$ in (23.74) are defined as follows:

$$\alpha(\omega) = \frac{1}{V} \sum_{\mathbf{k}} \frac{|\lambda(\mathbf{k})|^2}{\varepsilon(\mathbf{k})} \tanh \frac{\beta\varepsilon(\mathbf{k})}{2} \frac{\varepsilon^2(\mathbf{k}) + T^2(\mathbf{k}) + i\omega T(\mathbf{k})}{\omega^2 + 4\varepsilon^2(\mathbf{k})},$$

$$\beta(\omega) = \frac{\Delta^2}{V} \sum_{\mathbf{k}} \frac{|\lambda(\mathbf{k})|^4 \tanh \frac{\beta}{2} \varepsilon(\mathbf{k})}{\varepsilon(\mathbf{k})(\omega^2 + 4\varepsilon^2(\mathbf{k}))}. \quad (23.75)$$

There is a product of three factors on the right-hand side of (23.73). The first depends on V exponentially, the second is proportional to $V^{1/2}$ and the third is a constant as $V \to \infty$. The factors $e^{\pm i\omega\varepsilon}$ in front of $\alpha(\pm\omega)$ are necessary in order to obtain the correct value of the infinite product $\prod_{\omega > 0}$.

The problem of a rigorous justification of (23.70) and (23.73) for the BCS model is more difficult than the corresponding problem for the Dicke model. The point is that infinite products in the formulae for Z/Z_0 in the case of BCS model do not converge absolutely, while the corresponding products for the Dicke model are absolutely convergent.

That is why the technique of estimates developed for the Dicke model does not work in the BCS case where divergent infinite products arise. Nevertheless, the results for Z/Z_0 from (23.70) and (23.73) obtained for the BCS model seem to be valid. A conditional proof of these formulae based on an auxiliary hypothesis can be found in Popov (1980).

The final part of the section is devoted to the collective Bose excitations in the Dicke and BCS models.

For the Dicke model in the case $T > T_c$ we can approximate the effective action functional of the Bose field

$$-\pi\sum_{\omega} b^*(\omega)b(\omega) + N\,\mathrm{Tr}\ln(I + A) \tag{23.76}$$

by quadratic form

$$-\pi\sum_{\omega} b^*(\omega)b(\omega) + N\,\mathrm{Tr}\,BC = -\pi\sum_{\omega} b^*(\omega)b(\omega)(1 - a(\omega)), \tag{23.77}$$

where

$$a(\omega) = \frac{g^2 \tanh\frac{1}{4}\beta\Omega}{(\omega_0 - \mathrm{i}\omega)(\Omega - \mathrm{i}\omega)}. \tag{23.78}$$

The quadratic form (23.77) describes the one-mode Bose field. We may find its excitation spectrum from the equation

$$1 - a(\omega) = 0, \quad \mathrm{i}\omega \to E. \tag{23.79}$$

This equation can be rewritten as follows

$$E^2 - E(\Omega + \omega_0) + \omega_0\Omega - g^2 \tanh\tfrac{1}{4}\beta\Omega = 0, \tag{23.80}$$

which has two solutions

$$E_{1,2} = \tfrac{1}{2}(\Omega + \omega_0 \mp [(\Omega - \omega_0)^2 + 4g^2 \tanh\tfrac{1}{4}\beta\Omega]^{1/2}). \tag{23.81}$$

At the phase transition point $T = T_c$, we have $g^2 \tanh \beta_c\Omega/4 = \omega_0\Omega$ and we obtain

$$E_1 = 0, \quad E_2 = \Omega + \omega_0, \tag{23.82}$$

so the smallest solution of (23.80) vanishes.

Let us note that we could arrive at (23.79) from the condition that the factor $(1 - a(\omega))^{-1}$ in the finite product acquires a pole after the analytic condition $\mathrm{i}\omega \to E$. So it is sufficient to find Z/Z_0 in order to obtain information about the excitation spectrum. For example, we can calculate the excitation spectrum at $T < T_c$ using the equation

$$\det\begin{pmatrix} 1 - \alpha(\Delta, \omega), & \beta(\Delta, \omega) \\ \beta(\Delta, \omega), & 1 - \alpha(\Delta, -\omega) \end{pmatrix}$$
$$= 1 - \alpha(\Delta, \omega) - \alpha(\Delta, -\omega) + \alpha(\Delta, \omega)\alpha(\Delta, -\omega) - \beta^2(\Delta, \omega) = 0 \tag{23.83}$$

after changing $\mathrm{i}\omega$ to E. Substituting (23.50) into (23.83) we get the equation

$$E^2(E^2 - [(\Omega + \omega_0)^2 + 4g^2\Delta^2]) = 0, \tag{23.84}$$

which has roots

$$E_1 = 0, \quad E_2 = \pm[(\Omega + \omega_0)^2 + 4g^2\Delta^2]^{1/2}. \tag{23.85}$$

The zero solution $E_1 = 0$ corresponds to the Goldstone mode arising below T_c. Another approach for obtaining (23.85) was developed by Kirianov & Yarunin (1980).

The same method may be applied to the evaluation of the Bose spectrum of the BCS model. Equations for the Bose spectrum are

$$1 - a(\omega) = 1 - \frac{1}{V}\sum_{\mathbf{k}} \frac{|\lambda(\mathbf{k})|^2 \tanh \frac{1}{2}\beta T(\mathbf{k})}{2T(\mathbf{k}) - \mathrm{i}\omega} = 0, \quad T > T_c \tag{23.86}$$

$$\det\begin{pmatrix} 1 - \alpha(\omega), & \beta(\omega) \\ \beta(\omega), & 1 - \alpha(-\omega) \end{pmatrix}$$

$$= 1 - \alpha(\omega) - \alpha(-\omega) + \alpha(\omega)\alpha(-\omega) - \beta^2(\omega) = 0, \quad T < T_c. \tag{23.87}$$

The last equation for $T < T_c$ has a zero solution $\omega = 0$ (the Goldstone mode). This is clear from the identities

$$(1 - \alpha(0))^2 - \beta^2(0) = (1 - \alpha(0) - \beta(0))(1 - \alpha(0) + \beta(0)),$$

$$1 - \alpha(0) - \beta(0) = 1 - \frac{1}{2V}\sum_{\mathbf{k}} \frac{|\lambda(\mathbf{k})|^2}{\varepsilon(\mathbf{k})} \tanh \tfrac{1}{2}\beta\varepsilon(\mathbf{k}) = 0. \tag{23.88}$$

Here the last identity is just the gap equation (23.72).

The first equation in (23.86) has a zero solution $\omega = 0$, only at $T = T_c$. If $T > T_c, \omega \neq 0$. The second equation in (23.87) not only has the zero solution $\omega = 0$, but also a nonzero one.

It is easy to analyse the nonzero roots of (23.86) and (23.87) for the case where $\lambda(\mathbf{k}) = \lambda =$ constants in some narrow shell near the Fermi sphere $T(\mathbf{k}) = (k^2 - k_F^2)/2m = 0$. Here we may replace the sum by the integral over a neighbourhood of the Fermi sphere by making the substitution

$$V^{-1}\sum_{k} \to mk_F(2\pi)^{-3}\int \mathrm{d}\xi\,\mathrm{d}\Omega, \quad \xi = (k^2 - k_F^2)/2m. \tag{23.89}$$

The first equation in (23.86) may be rewritten as

$$\frac{mk_F\lambda^2}{2\pi^2}\int \frac{\mathrm{d}\xi \tanh \frac{1}{2}\beta\xi}{2\xi - \mathrm{i}\omega} = 1. \tag{23.90}$$

Using the equation for T_c

$$\frac{mk_F\lambda^2}{2\pi^2}\int \frac{\mathrm{d}\xi}{2\xi} \tanh \tfrac{1}{2}\beta_c\xi = 1 \tag{23.91}$$

we can write (23.90) as

$$\int \tanh \tfrac{1}{2}\beta\xi \left[\frac{1}{2\xi - \mathrm{i}\omega} - \frac{1}{2\xi}\right] \mathrm{d}\xi + \int \frac{\mathrm{d}\xi}{2\xi}(\tanh \tfrac{1}{2}\beta\xi - \tanh \tfrac{1}{2}\beta_c\xi) = 0, \quad (23.92)$$

or

$$\int \frac{\tanh \frac{1}{2}\beta\xi}{2\xi} \frac{\omega^2}{4\xi^2 + \omega^2} \mathrm{d}\xi + \int \frac{\mathrm{d}\xi}{2\xi}(\tanh \tfrac{1}{2}\beta\xi - \tanh \tfrac{1}{2}\beta_c\xi) = 0. \quad (23.93)$$

If $|T - T_c| \ll T_c, |\omega| \ll T = \beta^{-1}$, we have

$$\int \frac{\mathrm{d}\xi}{2\xi}(\tanh \tfrac{1}{2}\beta\xi - \tanh \tfrac{1}{2}\beta_c\xi) = \ln \frac{\beta}{\beta_c} \approx -\frac{T - T_c}{T_c}, \quad (23.94)$$

$$\int \frac{\tanh \frac{1}{2}\beta\xi}{2\xi} \frac{\omega^2}{4\xi^2 + \omega^2} \mathrm{d}\xi \approx \frac{\beta}{4} \int \frac{\omega^2 \mathrm{d}\xi}{4\xi^2 + \omega^2} = \frac{\pi\beta\omega}{8} \approx \frac{\pi\omega}{8T_c} \quad (23.95)$$

and (23.93) can be expressed as

$$-\frac{\pi\omega}{8T_c} - \frac{T - T_c}{T_c} = 0. \quad (23.96)$$

Replacing $\mathrm{i}\omega$ by E, we obtain

$$E = -\mathrm{i}\frac{8}{\pi}(T - T_c). \quad (23.97)$$

This result coincides with (13.17) for the Bose spectrum at $k = 0$ of a Fermi gas with attractive interaction.

Now let us consider a nonzero root of (23,87). After the substitution in (23.89) we may neglect the term

$$V^{-1} \sum_{\mathbf{k}} \frac{|\lambda(\mathbf{k})|^2}{\varepsilon(\mathbf{k})} \tanh \tfrac{1}{2}\beta\varepsilon(\mathbf{k}) \frac{\mathrm{i}\omega T(\mathbf{k})}{\omega^2 + 4\varepsilon^2(\mathbf{k})}$$

in $\alpha(\omega)$, see (23.75), and write (23.87) in the form

$$(1 - \alpha(\omega) - \beta(\omega))(1 - \alpha(\omega) + \beta(\omega)) = 0. \quad (23.98)$$

The first factor here vanishes at $\omega = 0$, as was shown above. Demanding that the second factor vanishes, we obtain

$$1 - \frac{\lambda^2}{V} \sum_{\mathbf{k}} \frac{1}{\varepsilon(\mathbf{k})} \tanh \tfrac{1}{2}\beta\varepsilon(\mathbf{k}) \frac{2T^2(\mathbf{k})}{\omega^2 + 4\varepsilon^2(\mathbf{k})} = 0 \quad (23.99)$$

$$\frac{\lambda^2}{V}\sum_{\mathbf{k}}\frac{1}{\varepsilon(\mathbf{k})}\tanh\tfrac{1}{2}\beta\varepsilon(\mathbf{k})\frac{\omega^2+4\lambda^2\Delta^2}{\omega^2+4\varepsilon^2(\mathbf{k})}=0, \tag{23.100}$$

where we have used the gap equation (23.72). This equation has the root

$$E=\pm 2\lambda\Delta. \tag{23.101}$$

We see that for all $T < T_c$ there exists a branch of the Bose spectrum equal to the double value of the gap in the Fermi spectrum. This result is in agreement with those obtained for the Bose spectrum of Fermi gas with attractive interaction (see (13.44)).

In conclusion we report the main results of this section.

(1) The quotient Z/Z_0 for the Dicke model is expressed via a functional integral with respect to a one-mode Bose field.

(2) The asymptotes of Z/Z_0 of large N are obtained. It is shown that one can give a rigorous proof for estimates (23.39), (23.48) and (23.52), which ensures the validity of the asymptotic formulae for Z/Z_0. Here we have outlined a rigorous proof only for the case $T > T_c + \delta$. Proofs for other cases ($T < T_c - \delta$, $T_c - \delta < T < T_c + \delta$) can be found in Popov & Fedotov (1981, 1982).

(3) Analogous asymptotes of Z/Z_0 are obtained for the BCS model. A conditional proof of these asymptotes based on an auxiliary conjecture can be found in Popov (1980).

(4) Equations for the Bose spectrum of the Dicke model and the BCS model are derived. It is shown that below T_c these equations have solutions corresponding to both the Goldstone mode ($E = 0$) and the non-Goldstone (gap) mode. Formulae obtained for the Bose spectrum of the BCS model at $T > T_c$ and for the gap mode of the model at $T < T_c$ are in agreement with those for the Bose spectrum of a Fermi gas with attractive interaction.

References

Abrikosov, A. A. (1952) *Dokl. Akad. Nauk SSSR* 86, 489

Abrikosov, A. A. (1957) *JETP* 32, 1442

Abrikosov, A. A. (1960) *JETP* 39, 1797–805

Abrikosov, A. A. (1961) *JETP* 41, 569–82

Abrikosov, A. A., Gor'kov, L. P. and Dzyaloshinki, I. E. (1962) *Methods of Quantum Field Theory in Statistical Physics.* Fizmatgiz, Moscow (in Russian)

Alonso, V. and Popov, V. N. (1977) *JETP* 77, 1445–59

Andrianov, V. A. and Popov, V. N. (1976) *Theor. and Math. Phys.* (*Sov.*) 28, 340–51

Andrianov, V. A., Kapitonov, V. S. and Popov, V. N. (1982) *Zap. Nauchn. Semin. LOMI* 120, 3–11

Andrianov, V. A., Kapitonov, V. S. and Popov, V. N. (1983) *Zap. Nauchn. Semin. LOMI* 131, 3–13

Andronikashvili, E. L. and Mamaladze, Yu. G. (1966) *Rev. Mod. Phys.* 38, 567–625

Anisimov, Yu. F. and Popov, V. N. (1985) *Zap. Nauchn. Semin. LOMI* 145, 22–33

Bardeen, J., Cooper, L. N. and Schriffer, J. R. (1957) *Phys. Rev.* 108, 1175–205

Beliaev, S. T. (1958*a*) *JETP* 34, 417–32

Beliaev, S. T. (1958*b*) *JETP* 2, 433–46

Berezin, F. A. (1966) *Method of Second Quantization.* Academic Press, NY

Berezinsky, V. L. (1970) *JETP* 59, 907

Berezinsky, V. L. (1971) *JETP* 61, 1144

Bidzhelova, G. K. and Popov, V. N. (1982) *Zap. Nauchn. Semin. LOMI* 120, 12–20

Bogoliubov, N. N. (1947) *Izv. Akad. Nauk. SSSR* 11, 77

Bogoliubov, N. N. (1960) *Physica* 26, 51

Bogoliubov, N. N. (1961) *Quasiaverages in Problems of Statistical Mechanics.* JINR Preprint D-781 (in Russian)

Bogoliubov, N. N. and Zubarev, D. N. (1955) *JETP* 28, 129–39

Bogoliubov, N. N., Tolmachev, V. V. and Shirkov, D. V. (1958) *New Method in Superconductivity Theory.* Academy of Sciences of the USSR, Moscow

Bogoliubov, N. N., Zubarev, D. N. and Tserkovnikov, Yu. A. (1960) *JETP* 39, 120

Bogoliubov, N. N. Jr. (1974) *Method of Investigation of Model Hamiltonians.* Nauka, Moscow (in Russian)

Bogoliubov, N. N. Jr. Brankov, I. G., Zagrebnov, V. A., Kurbatov, A. M. and Tonchev, N. S. (1981) *Method of Approximation Hamiltonians.* The Academy of Sciences of Bulgaria, Sofia

Brovman, E. G., Kagan Yu. M. and Holas, A. (1971) *JETP* 61, 2429–57

Brovman, E. G., Kagan, Yu. M. and Holas, A (1972) *JETP* 62, 1492–1501

Brusov, P. N. and Popov, V. N. (1980*a*) *JETP* 78, 234–45

Brusov, P. N. and Popov, V. N. (1980*b*) *JETP* 78, 2419–30

Brusov, P. N. and Popov, V. N. (1980*c*) 79, 1871–9

Brusov, P. N. and Popov, V. N. (1981) *JETP* 80, 1565–76

Brusov, P. N. and Popov, V. N. (1982*a*) *Phys. Lett.* 87A, 472–4

Brusov, P. N. and Popov, V. N. (1982*b*) *Theor. and Math. Phys.* (*Sov.*) 53, 444–55

Brusov, P. N. and Popov, V. N. (1984) *Collective Excitations in Superfluid Quantum Liquids.* Rostov University Press
Feynman, R. P. (1953) *Phys. Rev.* 91, 1921
Feynman, R. P. (1955) *Application of Quantum Mechanics to Liquid Helium.* Vol. 1. *Progress of Low Temperature Physics,* Chap. 2. North Holland, Amsterdam
Ginsburg, V. L. and Landau, L. D. (1950) *JETP* 20, 1064–82
Gombas, P. (1950) *Theorie und Lösungs Methoden des Mehrteilchenproblems der Wellenmechanik.* Basel
Hepp, E. and Lieb, E. H. (1973) *Ann. Phys.* 76, 360–404
Hugenholtz, N. M. and Pines, D. (1959) *Phys. Rev.* 116, 489–509
Kapitonov, V. S. (1978) *Zap. Nauchn. Semin. LOMI* 77, 84–105
Kapitonov, V. S. and Popov, V. N. (1972) *JETP* 63, 143–9
Kapitonov, V. S. and Popov, V. N. (1976) *Theor. and Math. Phys.* (*Sov.*) 26, 246–55
Kapitonov, V. S. and Popov, V. N. (1981) *Zap. Nauchn. Semin. LOMI* 101, 77–89
Khalatnikov, I. M. (1971) *Theory of Superfluidity.* Nauka, Moscow
Kirianov, V. B. and Yarunin, V. S. (1980) *Theor. and Math. Phys.* (*Sov.*) 43, 91–9
Kirjnitz, D. A. and Nepomniachij, Y. A. (1970) *JETP* 59, 2203–14
Kosterlitz, J. M. and Thouless, D. J. (1973) *J. Phys.* 66, 1181.
Landau, L. D. (1937) *JETP* 7, 627
Landau, L. D. (1941) *JETP* 11, 592
Landau, L. D. (1946) *JETP* 16, 574
Landau, L. D. (1956) *JETP* 30, 1058
Landau, L. D. (1957) *JETP* 32, 59
Lee, T. D. and Yang, C. N. (1958) *Phys. Rev.* 112, 1419, 1959
Lee, T. D. and Yang, C. N. (1959*a*) *Phys. Rev.* 113, 1165
Lee, T. D. and Yang, C. N. (1959*b*) *Phys. Rev.* 113, 1406
Lee, T. D. and Yang, C. N. (1959*c*) *Phys. Rev.* 116, 25
Lee, T. D. and Yang, C. N. (1960*a*) *Phys. Rev.* 117, 12
Lee, T. D. and Yang, C. N. (1960*b*) *Phys. Rev.* 117, 897
Luban, M. (1962) *Phys. Rev.* 128, 965–87
Matsubara, T. (1955) *Progr. Theor. Phys.* 14, 351
Moshchinsky, B. V. and Fedianin, V. K. (1977) *Theor. and Math. Phys.* (*Sov.*) 32, 96–100
Onsager, L. (1949) *Nuovo Cimento Sup.* 6, 279–87
Popov, V. N. (1964) *JETP* 47, 1759–64
Popov, V. N. (1965) *Vestn. Leningrad Univ.* 4, 58–64
Popov, V. N. (1971) *Theor. and Math. Phys.* (*Sov.*) 6, 90–108
Popov, V. N. (1972*a*) *Theor. and Math. Phys.* (*Sov.*) 11, 236–47
Popov, V. N. (1972*b*) *Theor, and Math. Phys.* (*Sov.*) 11, 354–65
Popov, V. N. (1973) *JETP* 64, 674–80
Popov, V. N. (1978) *Functional Integral in Quantum Field Theory and Statistical Mechanics.* Preprint Freie Universität Berlin West
Popov, V. N. (1980) *Zap. Nauchn. Semin. LOMI* 101, 128–50
Popov, V. N. and Fedotov, S. A. (1981) *Proceedings of the II International Symposium on the Selected Problems of Statistical Mechanics,* pp. 94–101
Popov, V. N. and Fedotov, S. A. (1982) *Theor. and Math. Phys.* (*Sov.*) 51, 73–85
Stein, D. L. and Cross, M. C. (1979) *Phys. Rev. Lett.* 42, 504–7
Svidzinski, A. V. (1971) *Theor. and Math. Phys.* (*Sov.*) 9, 273–90
Tolmachev, V. V. (1960*a*) *Dokl. Akad. Nauk SSSR* 134, 1324
Tolmachev, V. V. (1960*b*) *Dokl. Akad. Nauk SSSR* 135–41
Tolmachev, V. V. (1960*c*) *Dokl. Akad. Nauk SSSR* 135, 825–8
Tolmachev, V. V. (1969) *Theory of Bose Gas.* Moscow University Press
Tserkovnikov, Yu. A. (1962) *Dokl. Akad. Nauk SSSR* 143, 832
Tserkovnikov, Yu. A. (1964*a*) *Dokl. Akad. Nauk SSSR* 159, 1023
Tserkovnikov, Yu. A. (1964*b*) *Dokl. Akad. Nauk SSSR* 159, 1264

Suggested further reading

Andrianov, V. A. and Popov, V. N. (1974) *Vestn. Leningrad Univ.* no 16, 7–15

Andrianov, V. A. and Popov, V. N. (1976) *Vestn. Leningrad Univ.* no 22, 11–20

Bogoliubov, N. N. (1963) *On the Problem of Hydrodynamics of Superfluid Liquid.* JINR Preprint R-1491 (in Russian)

Bycling, E. (1964) *Ann. Phys.* 32, 367–76

Galasievich, Z. (1964) *Asymptotic Calculation of the Green Function in Viscose-Liquid Approximation for Superfluid Bose Systems.* JINR Preprint R-1517 (in Russian)

Gavoret, T. and Nozieres, P. (1964) *Ann. Phys.* 28, 349

Gell-Mann, M. and Brueckner, K. (1957) *Phys. Rev.* 106, 354–72

Girardeau, M. and Arnowitt, R. (1959) *Phys. Rev.* 113, 755

Huang, K. and Yang, C. N. (1957) *Phys. Rev.* 105, 767

Kapitonov, V. S. and Popov, V. N. (1972) *JETP* 63, 143–9

Lasher, G. (1968) *Phys. Rev.* 172, 224–9

Lieb, E. H. (1963) *Phys. Rev.* 130, 1616–24

Lieb, E. H. and Liniger, W. (1963) *Phys. Rev.* 130, 1605–16

Pitaevski, L. P. (1961) *JETP* 40, 646–51

Popov, V. N. (1983) *Functional Integrals in Quantum Field Theory and Statistical Physics.* Reidel Publishing Company

Popov, V. N. and Faddeev, L. D. (1964) *JETP* 47, 1315–21

Popov, V. N. and Yarunin, V. S. (1985) *Collective Effects in Quantum Statistics of Matter and Radiation.* Leningrad University Press

Reatto, L. and Chester, Q. V. (1967) *Phys. Rev.* 155, 88–100

Sonin, E. B. (1970) *JETP* 59, 1416–28

Yang, C. N. and Yang, C. P. (1969) *J. Math. Phys.* 10, 1115–22

Index

Printed in the United States
By Bookmasters